Workbook 5
Safety and Maintenance

Dr. Medhat Kamel Bahr Khalil, Ph.D, CFPHS, CFPAI.
Director of Professional Education and Research Development,
Applied Technology Center, Milwaukee School of Engineering,
Milwaukee, WI, USA.

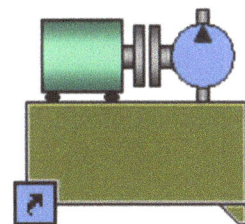

CompuDraulic LLC
www.CompuDraulic.com

CompuDraulic LLC

Workbook 5

Safety and Maintenance

ISBN: 978-0-9977816-6-3

Printed in the United States of America
First Published by XXXX
Revised by XXXX

Disclaimer

It is always advisable to review the relevant standards and the recommendations from the system manufacturer. However, the content of this book provides guidelines based on the author's experience.

Any portion of information presented in this book could be not applicable for some applications due to various reasons. Since errors can occur in circuits, tables, and text, the publisher assumes no liability for the safe and/or satisfactory operation of any system designed based on the information in this book.

The publisher does not endorse or recommend any brand name product by including such brand name products in this book. Conversely the publisher does not disapprove any brand name product by not including such brand name in this book. The publisher obtained data from catalogs, literatures, and material from hydraulic components and systems manufacturers based on their permissions. The publisher welcomes additional data from other sources for future editions.

Workbook 5
Safety and Maintenance
Table of Contents

PREFACE

This Workbook is a complementary part to the textbook of the same title. This book is used as a workbook for students to take notes during the course delivery. It contains colored printout of the PowerPoint slides that are designed to present the course. Each chapter is followed by a number of review questions and assignments for homework.

Dr. Medhat Kamel Bahr Khalil

Chapter 1
Hydraulic System Safety

Objectives:

Hydraulic equipment is widely used daily at almost all industrial sectors. Hydraulic power units range from a simple fraction of horsepower such as in a hydraulic hand tool to a large 500 horsepower machine. Lack of training, understanding how it works, and awareness of the associated hazards are reflected on increasing number of annual related injuries. The objective of this chapter is to increase the awareness about the hydraulic system safety during different phases including system design, startup, normal operation, and servicing. This chapter also explores the safety of the individuals, workplace, equipment, and the public.

0

0

Brief Contents:

1

1

1.1- Introduction

Please Note That:

- information provided in this book cannot cover every situation and is not intended to do so.

- It is highly recommended and always advisable to review the safety precautions provided by the components, systems, and machines manufacturers.

2

2

Electrical  =  Hydraulic

Fig. 1.1 – Safety of a Hydraulics System is as Important as that of an Electrical System

3

3

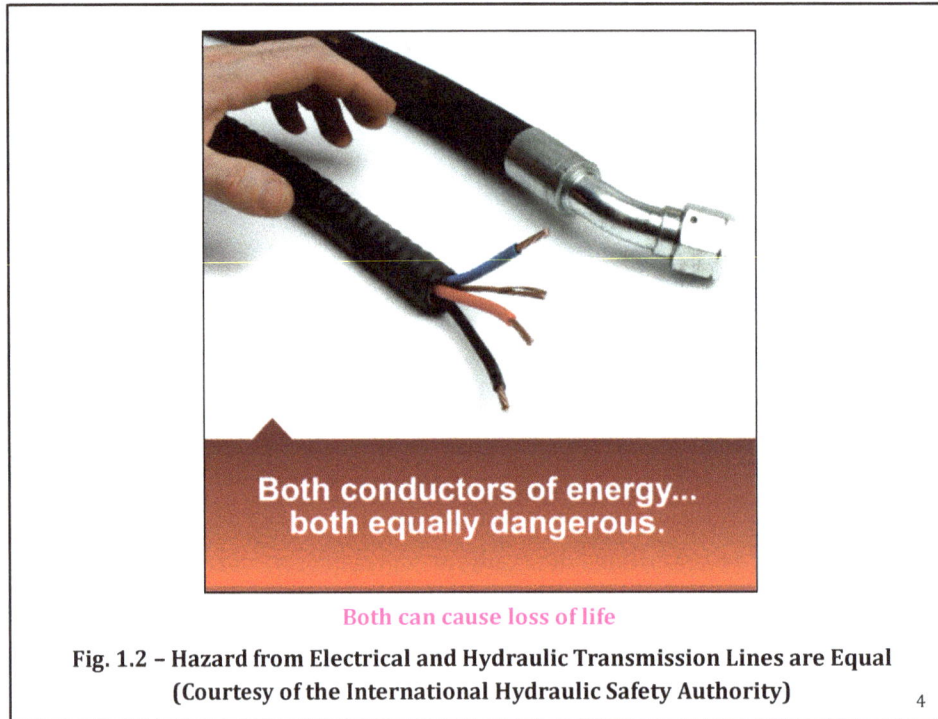

**Both conductors of energy...
both equally dangerous.**

Both can cause loss of life

**Fig. 1.2 – Hazard from Electrical and Hydraulic Transmission Lines are Equal
(Courtesy of the International Hydraulic Safety Authority)**

4

4

❏ **Why** hydraulic system safety is important?

❏ **Who** is responsible for the safety of hydraulic systems?

❏ **Where** to find general industry safety and health standards?

❏ **What** are the sources of best practices for hydraulic system safety?

❏ **When** to apply hydraulic system safety best practices?

Why?, Who?, Where? What? When?

Fig. 1.3 – Hydraulics System Safety Related Questions

5

5

1.2- Why Hydraulic System Safety is Important?

Video 276 (1 min)

"With the great power, there should be a great responsibility!!"

Fig. 1.4 – Hydraulic Power Associated with Large Loads

6

6

Why Hydraulic Systems Safety is Important?

V274 (2 min)

1. **Human:** injuries, fluid injection, loss of organs or life!. <u>Taking care of employees leverages productivity.</u>

2. **Environment:** oil spillage (cost and destruction)

3. **Surroundings:** fire, explosion, etc.

4. **Machine Itself:** damage + cost of shutdown.

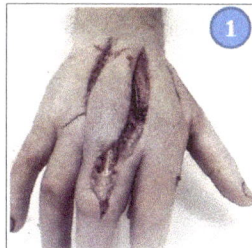

Fig. 1.5 – Why Hydraulics System Safety is Important?

7

7

1.3- Who is Responsible for Hydraulic Systems Safety?

Who is Responsible for the Hydraulic Systems Safety?

- **Managers.**
- **System Designers.**
- **Supervisors.**
- **Machine Operators.**
- **Servicing Personnel.**

I'm not responsible

I'm the only one responsible

Based on my job duty, I should follow pre-defined best practices

Fig. 1.6 – Who is Responsible for the Hydraulic System Safety?

8

8

1.4- Where to Find General Industry Safety and Health Standards?

Occupational Safety and Health Administration (OSHA):

Safety Standards #29 CFR 1910.
Publication # 2072:
General Industry Guide for Applying Safety and Health Standards:

❏ **Workplace Standards:**
Safety of floors, entrance and exit areas, sanitation, and fire protection.

❏ **Machines and Equipment Standards:**
- Machine guards + inspection + maintenance techniques.
- Safety devices + mounting of equipment.
- Noise levels produced by operating equipment.

❏ **Materials Standards:**
Toxic fumes, explosive dust particles, excessive atmospheric contamination.

9

9

❑ **Employee Standards:**
Training, personnel protective equipment, medical and first-aid services.

❑ **Power Source Standards:**
Power sources such as electric, hydraulic, pneumatic, and steam supply systems.

❑ **Process Standards:**
  Welding, spraying, abrasive blasting, and machining.

❑ **Administrative Regulations:**
Displaying of OSHA posters, stating the rights and responsibilities of both the employer and employee.

10

10

1.5- What are the Sources of Best Practices for Hydraulic System Safety?

Sources of Best Practices with Focus on Hydraulic System Safety

Manufacturers + Industry Standards + Organizations + Field Experts

Fig. 1.7 – Sources of Best Practices for Hydraulic System Safety

11

11

1.6- When to Apply Hydraulic System Safety Best Practices?

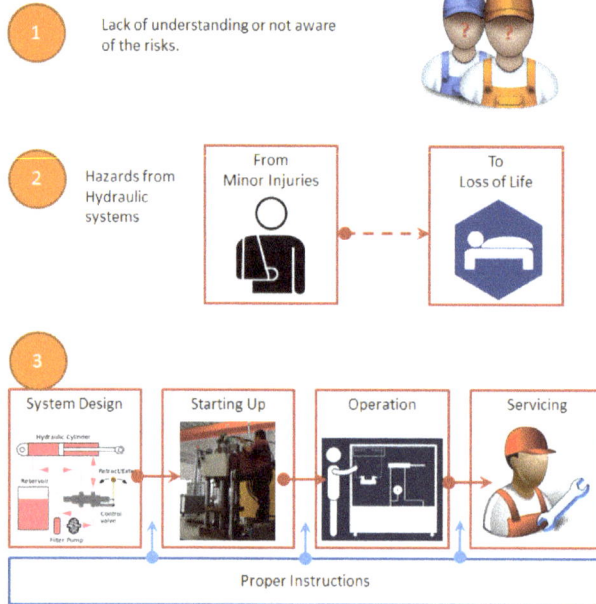

Fig. 1.8 – Challenges of Hydraulic System Safety

1. Lack of understanding or not aware of the risks.

2. Hazards from Hydraulic systems — From Minor Injuries → To Loss of Life

3. System Design → Starting Up → Operation → Servicing

Proper Instructions

12

Misconnection among these people can result in serious problems (Designer, Operator, and Servicer)

When to apply safety best practices?

During the following stages

1. System Design.
2. System Startup.
3. System Operation.
4. System Servicing.

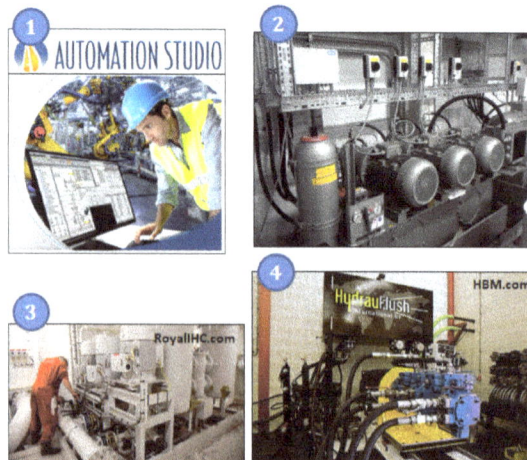

Fig. 1.9 – When to Apply Hydraulic System Safety Best Practices?

13

This Chapter Provides Detailed Interpretation for:

❑ **BP-Safety-01:** Design for Safe and Reliable Hydraulic Systems.

❑ **BP-Safety-02:** Safety of Hydraulic System Operators.

❑ **BP-Safety-03:** Safety of Hydraulic System Work Environment.

❑ **BP-Safety-04:** Safety of Hydraulic System Workspace.

❑ **BP-Safety-05:** Safe Startup of Hydraulic Systems.

❑ **BP-Safety-06:** Safe Operation of Hydraulic Systems.

❑ **BP-Safety-07:** Safe Servicing of Hydraulic Systems.

❑ **BP-Safety-08:** Oil Injection Avoidance and Treatment.

❑ **BP-Safety-09:** Safe usage of Hydraulic Powered Tools.

Please Note That:

▪ The set of the Best Practices provided in this chapter are not intend to replace the approved industry standards, it rather provides the author's point of view and personal experience.

14

14

1.7- BP-Safety-01: Design for Safe and Reliable Hydraulic Systems

1. Review Manufacturers' Recommendations.
2. Work with Standards.
3. Limit Maximum Operating Pressure.
4. Limit Maximum Operating Temperature.
5. Provide Protection Against Surface Temperature.
6. Eliminate Risk of Fire and Explosion.
7. Adequately Size and Select Hydraulic Transmission Lines.
8. Adequately Size and Select Hydraulic Components.
9. Avoid Pump Cavitation.
10. Carefully Design Hydraulic Fluid Contamination Control Systems.
11. Minimize Noise and Vibrations of the System.
12. Apply Energy Saving Design Strategies for Hydraulic Systems.
13. Apply Fail-Safe Design Strategies for Hydraulic Systems.
14. Properly Design Condition Monitoring System.
15. Perform Dynamic Analysis Whenever Needed.

15

15

1.7.1- Review Manufacturers' Recommendations

- System Designer **MUST** review Manufacturer's Recommendations.

- Components' Manufacturers → Limits (min & max) for Operating Conditions.

- Disrespect of Manufacturer's Recommendations →

    o Unreliable System.

    o Accidental Machine Breakage.

    o Possible Load Uncontrolled.

    o Unsafe Work Environment.

16

16

- **Example 1:**
- Exceeding maximum viscosity → pump cavitation and overall system failure.

Variable Axial Piston Pump
(A)A10VSO

Viscosity limits

The limiting values for viscosity are as follows:

V_{min} = 60 SUS (10 mm²/s)
short term (t ≤ 1 min)
at a max. permissible leakage oil temperature
of t_{max} = 195 °F (90 °C).

Please note that the max. fluid temperature of 195 °F (90 °C) is
also not exceeded in certain areas (for instance bearing area).
The temperature in the bearing area is approx. 7 °F (5 K)
higher than the average leakage fluid temperature.

V_{max} = 7500 SUS (1600 mm²/s)
short term (t ≤ 1min)
on cold start
(p ≤ 435 psi/30 bar, n ≤ 1000 rpm, t_{min} = -13 °F/-25 °C)

**Fig. 1.10 – Operating Conditions Specified by a Pumps Manufacturer
(Courtesy of Bosch Rexroth)**

17

17

- **Example 2:**
- Exceeding maximum pressure → hose blowing out.
- Exceeding minimum bend radius → more stress on reinforcement layer + outer protective surface hose cracked.

V277 (1 min)

I.D. INCHES	DASH NO. REF.	SAE NO. & TYPE SPEC.	O.D. MAX. INCHES	MIN. BEND (INTERNAL) RAD. IN. AT MAX. OPERATING PRESSURE	¹MAX. OPERATING PRESSURE PSIG
1/8	-3	100R14	0.268	1.5	1,500
3/16	-3	100R1-A	0.531	3.5	3,000
3/16	-3	100R1-AT	0.494	3.5	3,000
3/16	-3	100R2-A&B	0.656	3.5	5,000

¹Minimum burst pressure is 4 times maximum operating pressure.

Fig. 1.11 – Operating Conditions Specified by a Hose Manufacturer

18

1.7.2- Work with Standards

Working on your own making inexperienced estimates and decisions may result in unsafe machines.

❑ **LRHB:**
- Standard engineering data + rules of thump.

❑ **International Standard Organization (ISO)**
- Graphics and Symbols Standards **(ISO 1219).**
- Terms and definitions **(ISO 5598).**

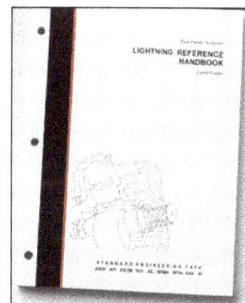

Fig. 1.12 – Lightening Reference Handbook

○ **Design Standards**
  ○ ISO 4413 (NFPA T2.24.1) - General Req. for Safety of Hydraulic Systems.
  ○ ISO 4414 (NFPA T2.24.2) - General Req. for Safety of Pneumatic Systems.
  ○ ISO 13849-1 – Safety of Control Systems.

❑ **Performance Standards**
  ○ Pressure Rating.
  ○ Contamination Control.
  ○ Performance Standard Tests.

19

More Background in the Textbook

Source of Standards	Year Founded	Website
Society of Automotive Engineers (SAE).	1905	www.sae.org
Joint Industry Conference (JIC).	1950	-
National Fluid Power Association (NFPA).	1953	www.nfpa.com
International Fluid Power Society (FPS).	1960	www.ifps.org
American National Standards Institute (ANSI) B93	1961	www.ansi.org
International Standards Organization (ISO) TC131.	1970	http://www.iso.org

Table 1.1 - Standards that Contain Information about Fluid Power Systems

20

1.7.3- Limit Maximum Operating Pressure

- Working Pressure:
  - ○ Steady Load → Steady Pressure.
  - ○ Dynamic Loads → Pulsating Pressure.
  - ○ Shock loads → Pressure Spikes.
  - ○ Differential Cylinders → Pressure Intensification.
  - ○ Thermal expansion → Pressure Intensification.
- Noncompliance with the P_{max}. → Unsafe Machine Operation.

- **Example 1:**

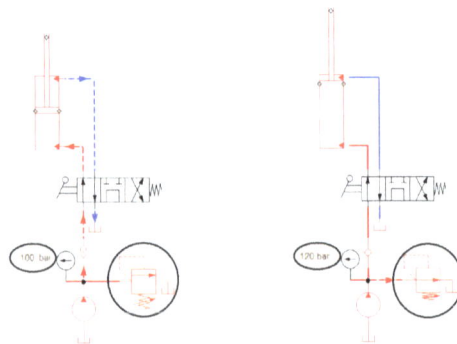

Fig. 1.13 – Limit Maximum Pressure at Pump Outlet

21

- **Example 2:**

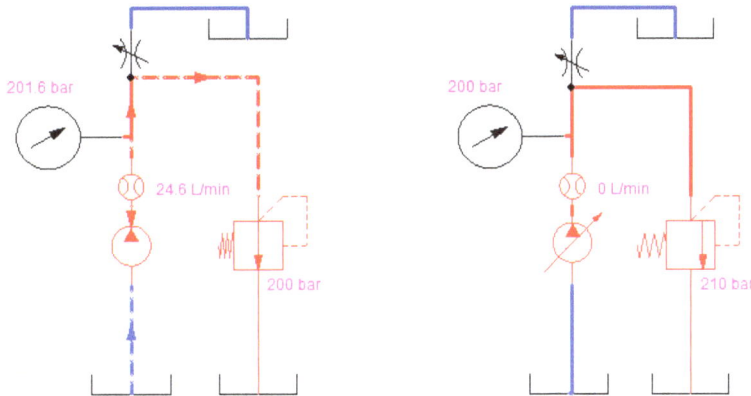

201.6 bar

24.6 L/min

200 bar

200 bar

0 L/min

210 bar

Fig. 1.14 – Pressure Compensated Pump Limits Maximum Pressure with Energy Saving

22

22

- **Example 3:**

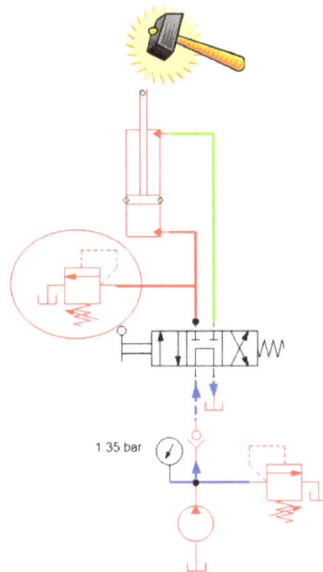

1.35 bar

Fig. 1.15 – A Secondary Relief Valve Protects other Components in the System

23

23

- **Example 4:**

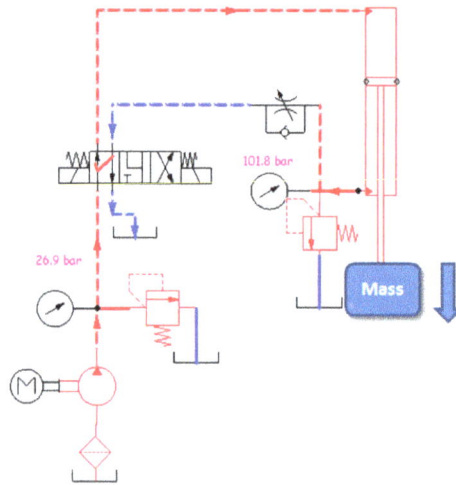

Fig. 1.16 – A Secondary Relief Valve Limits Pressure Intensification in a Differential Cylinder

24

24

- **Example 5:**

- low flow thermal PRV
- Preventing cylinder damage from thermal pressure intensification at port 2.
- Not for use in dynamic pressure limiting applications.

CV10-28 — Check Valve with Thermal Relief

Thermal Relief Settings:
05	34.5 – 48.3 bar	(500 – 700 psi)
10	69.0 – 93.1 bar	(1000 – 1350 psi)
20	137.9 – 172.4 bar	(2000 – 2500 psi)
25	172.4 – 217.2 bar	(2500 – 3150 psi)
30	206.9 – 262.1 bar	(3000 – 3800 psi)
40	275.9 – 344.8 bar	(4000 – 5000 psi)
45	310.3 – 386.2 bar	(4500 – 5600 psi)

SYMBOL:

Fig. 1.17 – A Special Relief Valve Limits Pressure Increase due to Thermal Expansion (Courtesy of Hydraforce)

25

25

- **Example 6:**

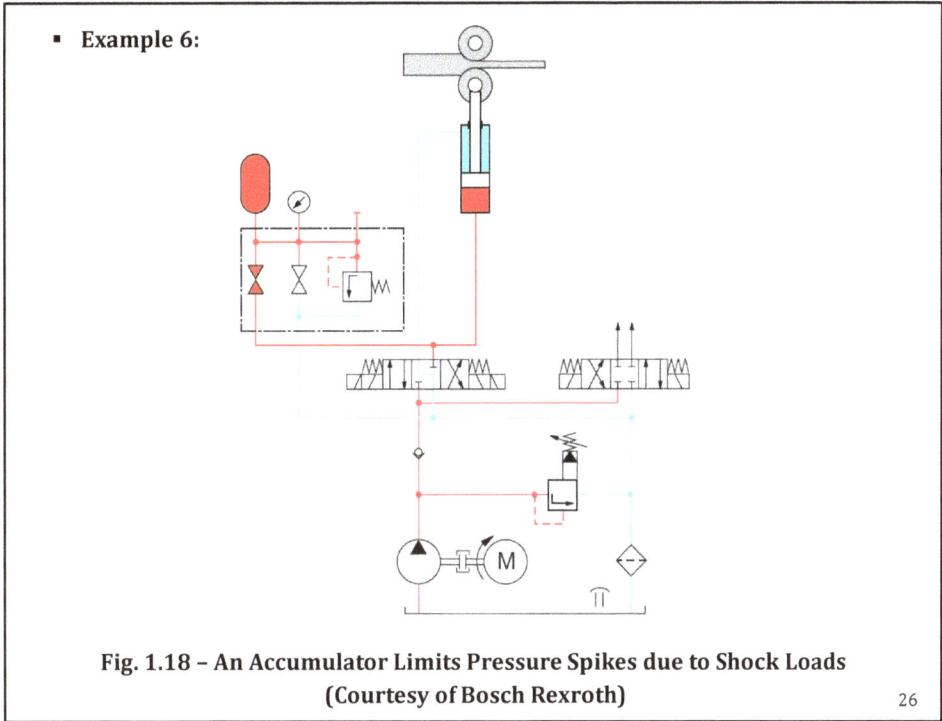

**Fig. 1.18 – An Accumulator Limits Pressure Spikes due to Shock Loads
(Courtesy of Bosch Rexroth)**

26

26

1.7.4- Limit Maximum Operating Temperature

- Hydraulic fluid is the life blood of the machine.

- Fluid properties are highly affected by working Temp

- Noncompliance with the T_{max}. $\rightarrow$ Unsafe Machine Operation.

- Heat balance equation.

Fig. 1.19 – Solving Heat Balance Equation for Proper Sizing a Heat Exchanger

Heat Imported (1) + Heat Generated (2) – Heat dissipated (3)
= Size of Heat Exchange (4)

27

27

1.7.5- Provide Protection against Surface Temperature

Burns from Hot Surfaces: hand burn as a result of accidental touching a hot surface of a hydraulic machine.

Burns from Conveyed Fluids: Contact with hot fluid, especially at temperatures over 60 $^{\circ}$C [140 $^{\circ}$F], can burn human skin.

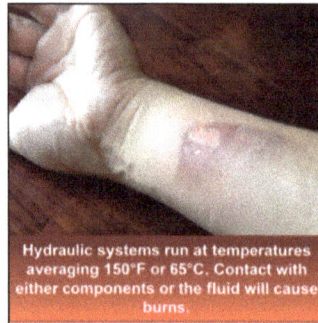

Hydraulic systems run at temperatures averaging 150°F or 65°C. Contact with either components or the fluid will cause burns.

**Fig. 1.20 – Hazard from Touching Hot Surfaces of a Hydraulic Machine
(Courtesy of the International Hydraulic Safety Authority)**

28

28

Design Solutions:
- Systems should be designed to protect against Surface Temperatures.
- Hose guards and pipe shields.
- Placing the hot surface in an inaccessible location.
- Proper warnings shall be provided.

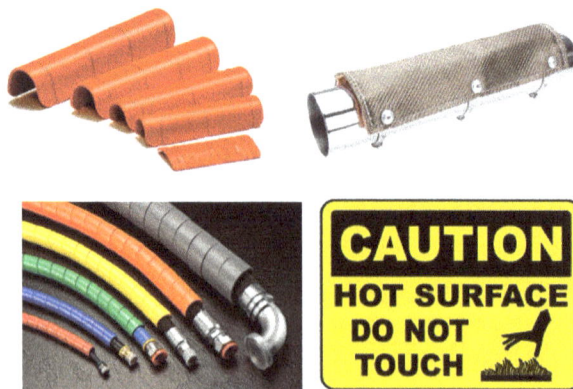

CAUTION
HOT SURFACE
DO NOT TOUCH

Fig. 1.21 – Protection against Surface Temperature

29

29

1.7.6- Eliminate Risk of Fire and Explosion

Source 1: Hydraulic fluids

Solutions:
- Use area guards and hose shields.

- Use fire-resistant fluids for applications associated with high heat sources.

V279 (4 min)

**Fig. 1.22 – Hazard from Burning Hydraulic Fluid
(Courtesy of the International Hydraulic Safety Authority)**

30

30

Source 2:
- Static eclectic discharges in Lines and Reservoirs.
- Hydraulic systems located near high voltage electrical power lines.

Solutions:
- Use nonconductive hoses (Orange Color) near electrical power lines.
- Properly ground the metallic reservoir or isolate it by rubber basses.
- Use plastic or fiberglass type reservoirs for mobile applications.

Non-Conductive high density fiber braid reinforced synthetic rubber hose for use in hydraulic systems where high voltage is present.

V146 (1 min)

Fig. 1.23 – Solutions to avoid Electric Static Discharge

31

31

Source 3: potential sources of ignition

Solution:
Locate or shield potential sources of ignition away from hydraulic equipment.

Such as:
- Electric motors.
- Hot surfaces
- Open flames

32

32

1.7.7- Adequately Size and Select Hydraulic Transmission Lines

A system designer must be aware of:

- Different methods of sizing a hydraulic transmission line (rules of thumb, equations, charts, and software).

- BP of Selecting a type of the hydraulic line (pipe, tube, and hoses) depends on the line size, length, application, and pressure rating.

- Trusted sources. V283 (4.5 min)

- ISO3457: the operators must be protected from any hose assembly located within 3 feet.

Low pressure adapters in a high pressure system is extremely hazardous.

Fig. 1.24.A – Hazard from Improper Selection of a Hydraulic Transmission Line (Courtesy of the International Hydraulic Safety Authority)

Fig. 1.24.B- Protective Hose Sleeve (Courtesy of Gates)

33

33

1.7.8- Adequately Size and Select Hydraulic Components

- Components sized based on the maximum expected flow.

- Under-sizing valves → act like throttles regardless the basic valve function,

- Oversizing valves → reduce their controllability.

- Wrong assumption (the maximum flow in the circuit is the pump flow!)

- Maximum flow in the circuit exceeds the pump flow in some cases, e.g.:
  o During retracting a differential cylinder, such as in cycling a cylinder in machine tools.
  o Flow surges after releasing high-pressure acting on a large volume of oil such, as in hydraulic presses.
  o Flow surges due to overrunning loads, such as in earth moving machines.
  o Flow surges from accumulators, such as in die-casting and injection molding machines.

34

34

1.7.9- Avoid Pump Cavitation

Pump cavitation is a major cause of pump failure and consequently the rest of the system. A system designer must be aware of the best practices of a hydraulic system design, installation, and operation to eliminate possibilities of developing cavitation. These best practices will be discussed in the next volume in this textbook series and includes:

- Reservoir design and placement.
- Pump intake line sizing and installation.
- Protection instruments to monitor the vacuum conditions at the pump intake.
- Fluid selection.
- Pump driving speed.
- Maintaining recommended working temperature.

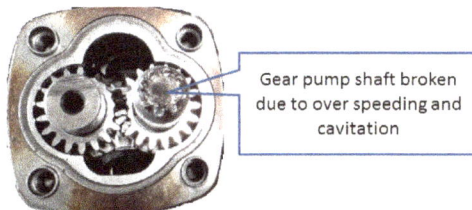

Gear pump shaft broken due to over speeding and cavitation

Fig. 1.25 – Hazards from Pump Cavitation

V280 (1 min)

35

35

1.7.10- Carefully Design Hydraulic Fluid Contamination Control Systems

- 80% of the failure is due to contamination.
- Hazards from hydraulic fluid contamination are not limited to just wearing a components, it could result in a disaster.
- A hydraulic system designer must be aware of the best practices for controlling the contamination of a hydraulic system that is discussed in the chapter 4 in this textbook.

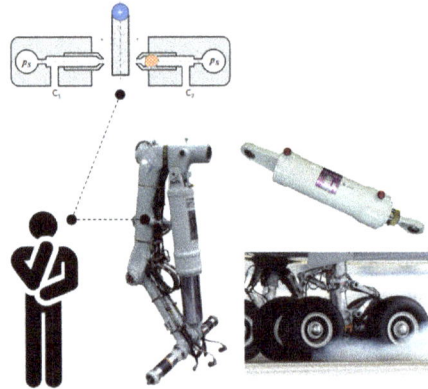

Fig. 1.26 – Hazards from Hydraulic Fluid Contamination

36

36

1.7.11- Minimize Noise and Vibration of the System

- Vibration $\to$ fatigue failure + loosens fittings.

- Noise $\to$ long-term personal disability.

- Structure-borne, fluid-borne, and Airborne noise are due to many things such as:
  - Pump-motor installation.
  - Pressure ripples.
  - Layout of transmission lines.
  - Rotating parts unbalance and not covered.

- Best practices (out of the scope of this textbook).

37

37

Example 1

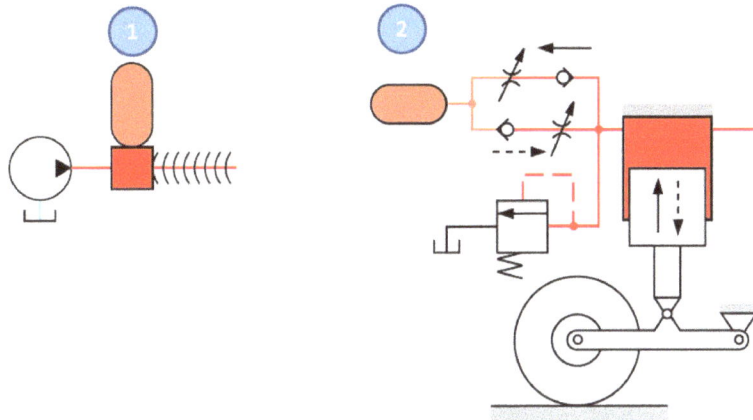

**Fig. 1.27 – Using an Accumulator to Minimize Noise and Vibration
(Courtesy of Bosch Rexroth)**

38

Example 2

- An inner radial chamber with a series of 0.5-in. diameter holes.
- A compressed coil spring surrounding the inner chamber.
- An outer radial chamber dotted with 0.03in. holes.

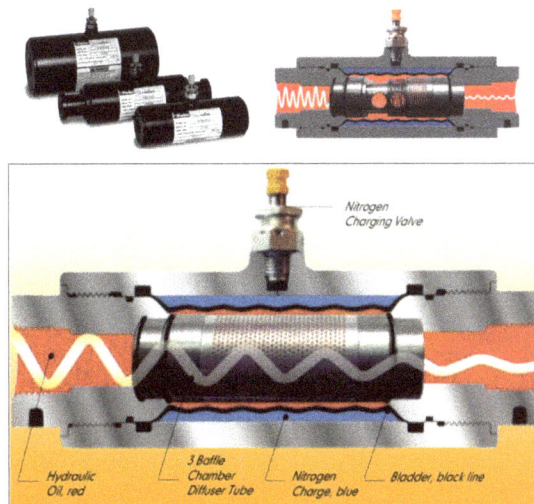

**Fig. 1.28 – Using a Shock
Suppressor to Minimize
Noise and Vibration
(Courtesy of Parker)**

39

1.7.12- Apply Energy Saving Design Strategies for Hydraulic Systems

- Energy waste → heat added to the system.
- Overheated system → system failure.
- "Energy Saving Design Strategies" (out of the scope of this textbook).
- Simple Example:

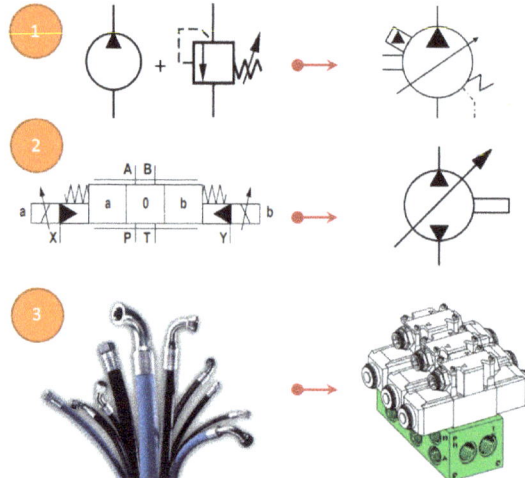

Fig. 1.29 – Examples of Energy Saving Solutions

40

40

1.7.13- Apply Fail-Safe Design Strategies for Hydraulic Systems

- Think of how to make a system fail safely.
- Consider all possible modes of system failure.
- " Fail-Safe Design Strategies" (out of the scope of this textbook).

Example 1 - Emergency Stop:
- Readily accessible.
- must be energized independently.

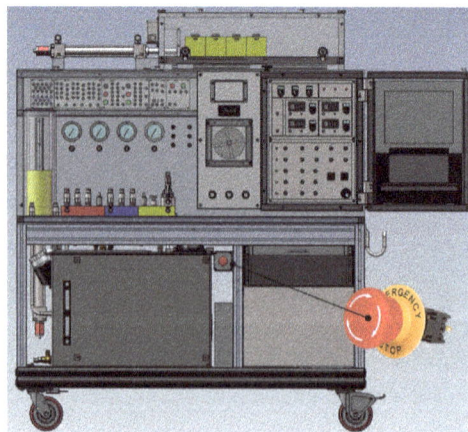

Fig. 1.30 – Emergency Stops

41

41

- **Example 2 – Manual Override for Solenoid-Operated Directional Valves:**
  - Spool can still be shifted in case of power outage or loss of signal.

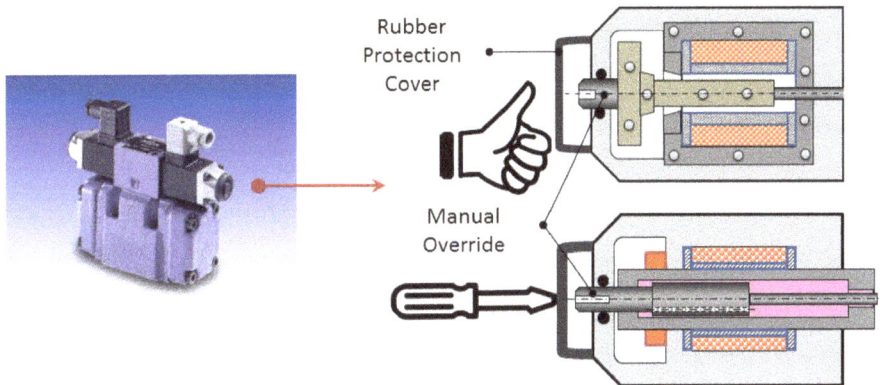

Fig. 1.31 – Manual Override for Solenoid-Operated Hydraulic Directional Valves

42

42

- **Example 3 – Emergency Control:**
  - Separate or part of the original system.
  - Drives the system to a safe condition (lock, complete cycle, return).
  - Triggered automatically in case of emergency.

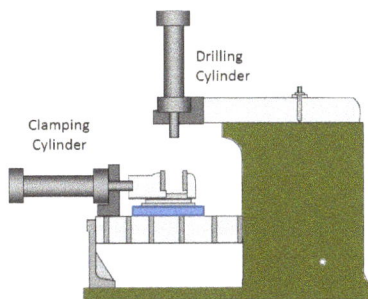

Fig. 1.32 – An Example of Emergency Control Concept in a Machine Tool

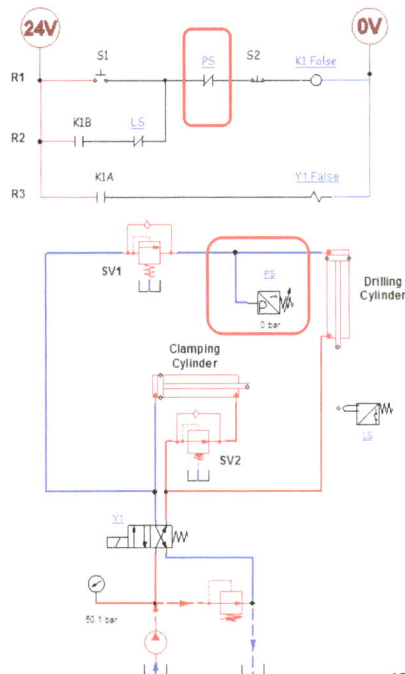

43

43

- **Example 4 – Automatic Release of Stored Energy:**

  o Stored energy must be released in case of emergency of machine shutdown.

  o If the accumulator is vented manually by a bleed valve, complete information should be given on or near the accumulator as well as on the circuit diagram.

  o Review **OSHA Standard 3120** for proper labeling/tagout of stored energy equipment.

  Fig. 1.33 – Automatic Release of Stored Energy

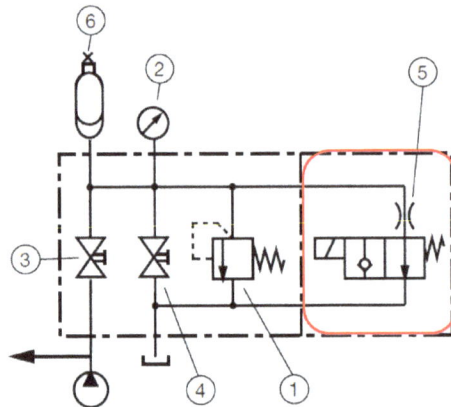

Schematic of a HYDAC Safety and Shut-off Block

1 - pressure relief valve

2 - pressure gauge *(optional)*

3 - shut-off valve

4 - manual bleed valve

5 - solenoid operated bleed valve *(optional)*

6 - thermal fuse cap *(optional)*

44

44

- **Example 5 – Secure Overrunning load:**
  o Float-Center type DCV.
  o Valve built in with the cylinder.
  o Pressure intensification at the rod side must be considered.
  o If no choice but using a line between the valve and the cylinder, DO NOT use flexible hose.

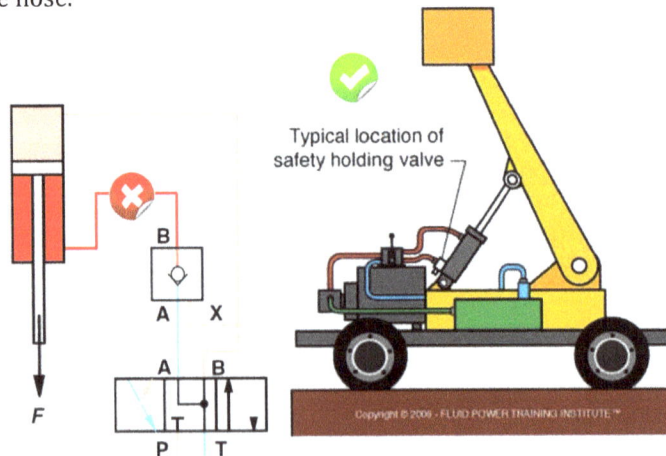

Typical location of safety holding valve

Copyright © 2009 - FLUID POWER TRAINING INSTITUTE ™

Fig. 1.34 – Securing Overrunning Load   45

45

1.7.14 - Properly Design Condition Monitoring System

- Hydraulic systems are non-transparent.
- Working conditions (pressure, temperature, flow, oil level, linear position, linear speed, acceleration, RPM, vibration, sound level, etc.)
- Gauges or Sensors.

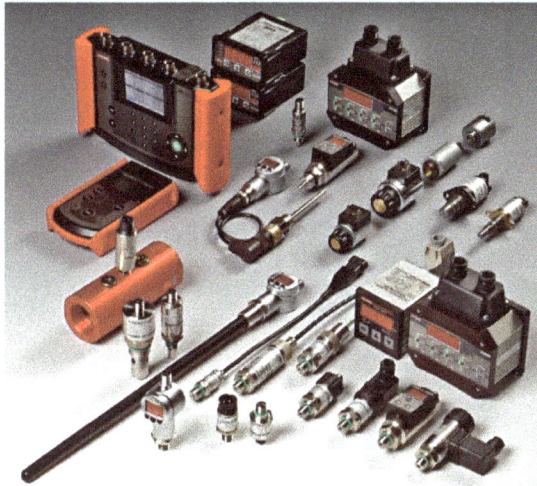

Fig. 1.35 – Hydraulic System Condition Monitoring Devices (Courtesy of Hydac)

46

46

1.7.15 – Perform Dynamic Analysis Whenever Needed

- Time-invariant (*Steady State*) equations my not be sufficient to predict system response.

- Dynamic analysis, modeling and simulation to avoid unsafe and unexpected system responses.

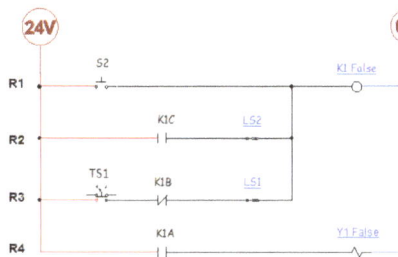

Example 1: Pressure Spikes in Highly Dynamic Systems

Fig. 1.36 – Pressure Spikes in a Cylinder Reciprocation System

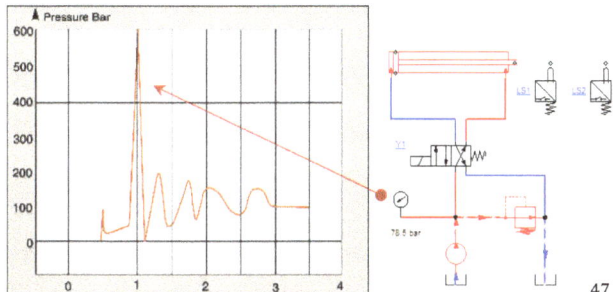

47

47

Example 2: Effect of Oil Compressibility in High Pressure Applications

when the oil is depressurized to retract the ram, the oil will expand suddenly causing surge of flow and damage the control valves.

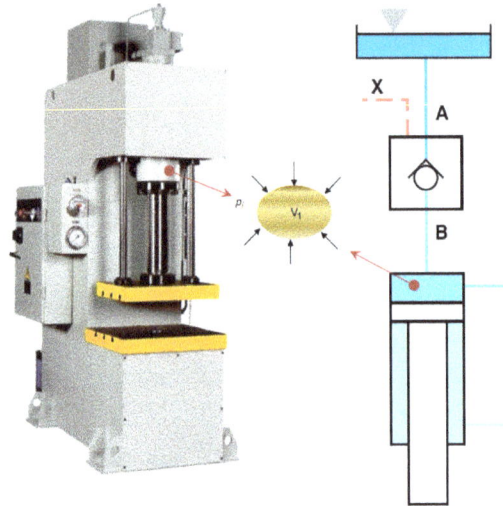

Fig. 1.37 – Effect of Oil Compressibility in High Pressure Applications

48

48

1.8- BP-Safety-02: Safety of Hydraulic System Operators

Safe operation of hydraulic system begins before work starts

**Designing for safe hydraulic systems
is not the end of the story!**

**Operating a hydraulic system must guarantee the safety of the
operator, work environment, and workspace.**

Safety of the operator comes first

PB-Safety 02: Safety of Hydraulic System Operators
1. Eye Protection.
2. Ear Protection.
3. Hand Protection.
4. Foot Protection.
5. Head Protection.
6. General Body Protection.

49

49

1.8.1- Eye Protection

- Oil leaks can occur without warning.
- Operators must wear proper Eye Protection devices,
- Particularly when disconnecting transmission lines.
- Various types: plastic goggles to face shields.
- Eye Washer stations.

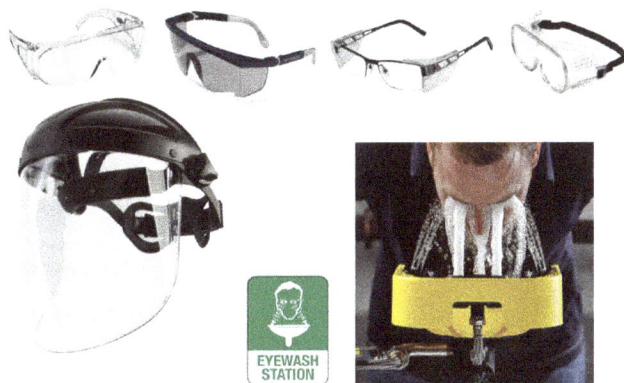

Fig. 1.38 – Eye Protection for Hydraulic System Operator

50

50

1.8.2- Ear Protection

Earplugs: most popular, inexpensive, and very common. Made from plastic or foam.

Earmuffs: more expensive, more effective, and more comfortable than earplugs.

Helmets: most expensive. Usually used only in the most severe noise conditions or where a combination of a hardhat and hearing protection is required.

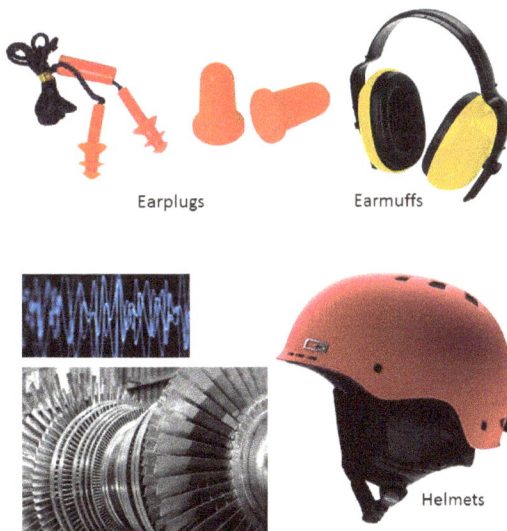

Earplugs   Earmuffs

Helmets

Fig. 1.39 – Ear Protection for Hydraulic System Operator

51

51

1.8.3- Hand Protection

- Oil leaks can occur without warning.
- Operators must wear special industrial gloves against oil injection.
- Particularly when disconnecting transmission lines.
- Continual skin contact with hydraulic fluid is to be avoided.
- Careful skin cleaning of sticky fluid is required.

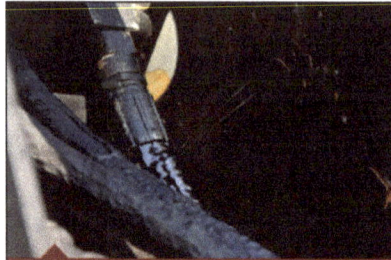

The high pressure jet of fluid from a pinhole rupture is extremely dangerous. Contact would cause a fluid injection injury which could lead to amputation.

Fig. 1.40 – Jet of Fluid at High Pressure from a Pinhole (Courtesy of the International Hydraulic Safety Authority)

Fig. 1.41 – Industrial Gloves against Oil Injection

52

52

1.8.4- Foot Protection

Hydraulic components are heavy and can cause injuries if they fall down accidently. Therefore, hydraulic system operators must wear approved safety shoes. Open footwear is forbidden in the work area.

Fig. 1.42 – Industrial Safety Shoes

53

53

1.8.5- Head Protection

- A Safety helmet (hard hats) protect the head from injuries caused by falling or moving objects.
- Such hard hats are generally made from plastic.
- Provide limited protection from heat and electrical shock.
- It can also be equipped with ear and eye protection.

Fig. 1.43 – Industrial Gloves against Oil Injection

54

54

1.8.6- General Body Protection

The following are general guidelines for body protection:

- Remove ties, rings, watches and jewelry.
- Most hydraulic fluids are toxic. No food in the work area.
- Hydraulic fluids can become HOT! Be aware of burn hazards.
- Avoid being in a wrong position.
- DO NOT expose yourself to actuator's movement.
- DO NOT stand under hydraulic supported equipment unless it is mechanically locked in place.

Fig. 1.44 – Avoid Being in Wrong Position
(Courtesy of Fluid Power Training. Inc.)

55

55

- Hyd. actuators generate high forces. DO NOT underestimate the load!

Reported Case History:
Action: an assembly-line worker who was installing a hydraulic steering cylinder on a front-end loader.

Result: An accident claimed four of his fingers.

Reason: Improper air bleeding. Worker was not aware of crushing hazard. Cylinder moves erratic pinning his hand in-between the rod-eye and the frame anchor.

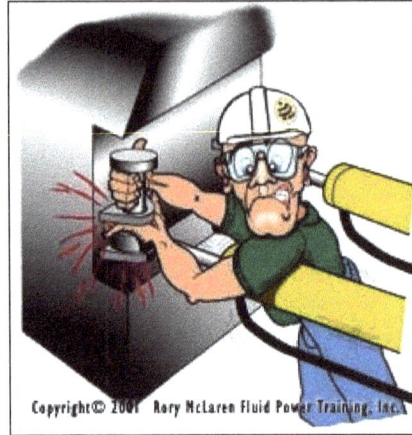

Copyright© 2001 Rory McLaren Fluid Power Training, Inc.

Fig. 1.45 – Cylinder Crushing Operator's Hand (Courtesy of Fluid Power Training. Inc.)

56

56

1.9- BP-Safety-03: Safety of Hydraulic System Work Environment

- Take care of your employees; they will take care of you.

- Improving the work environment → increases the comfort level of employees → increases productivity.

- Controlling the work environment → prevents cumulative injuries.

- *Cumulative injuries* are those injuries that occur from long-term exposure to unsafe environmental conditions.

- Work environmental conditions must meet or exceed state and federal standards, including those of the Occupational Safety and Health Administration (OSHA).

  PB-Safety 03:
  1. Air Quality.
  2. Ambient Conditions.
  3. Light Level.
  4. Sound Level.

57

57

1.9.1- Air Quality

- Ventilation and air circulation all time.

- Fumes are good sources for health issues.

Fig. 1.46 – Fumes at Steel Works

58

58

1.9.2- Ambient Conditions

- Controlling ambient temperature and humidity.
- → improving attention level during the duration of work.
- → reduce accidents.

1.9.3- Level of Light

- **OSHA standard 1926.56** → minimum lighting requirements = 10 (ftc).
- Unit of (ftc) "foot-candle".
- (ftc) is the illumination produced by a candle from 1 foot distance.
- OSHA → all emergency exit routes must be illuminated and marked "Exit.".
- Emergency lighting must function even in case of power failure.

59

59

1.9.4- Level of Sound

- **OSHA (Act of 1970)** → permissible noise exposure.

- Sound level in work environment can be reduced by:
  - o Sound absorbent materials on walls, ceilings, and partitions.
  - o Sound-reducing enclosures around devices/machines.
  - o If no choice, ear protection must be provided.

Hours/Day	Sound Level (db.)
8	90
6	92
4	95
3	97
2	100
1-1.5	102
1	105
0.5	110
0.25 or less	115

Table 1.2– Permissible Noise Exposure

60

60

1.10- BP-Safety-04: Safety of Hydraulic System Workspace

To ensure safety of hydraulic systems operation, some safety regulations must be applied to the workspace. BP-Safety-04 presents guideline for the safety of a hydraulic systems workspace

PB-Safety 04:
1. General Cleanliness and Organization.
2. OSHA Floor Marking.
3. Medical First Aid.
4. Fire Fighting Equipment.
5. Secured Hazardous Areas.
6. Safety and Job Performance Posters.

V282 (1 min)

The following subsections provide detailed interpretation of the action items listed in BP-Safety-05.

61

61

1.10.1- General Cleanliness and Organization

General housekeeping has a great influence on the number of accidents and injuries in a hydraulic system workspace.

Workplace must be well organized and cleared from all unsafe conditions such as:

- Open flam or welding spots.
- Environmental temperature higher than (150 ^{0}F or 66 ^{0}C).
- Dangerous electrical connections.
- Trip hazards.
- Spilled liquids.
- Standing water.

Concrete floors as foundations should be protected against fluids by being sealed or being painted with fluid-resistant paint.

62

62

V265- (2 min)

Fig. 1.47 – Clean and Organized Workspace

63

63

1.10.2- OSHA Floor Marking

OSHA Standard 1910.22 Floor Marking Standard.
Maintain clear egress pathways at least 24 inches wide at all times.

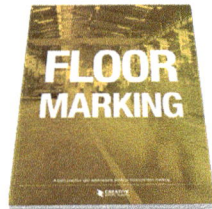

Fig. 1.48 –Workspace Floor Marking

64

64

1.10.3- Medical First Aid

- Work injuries can happen even in an organized workspace.

- Rapid response to such work injuries is very important to minimize the consequences of the injuries.

- Industrial medical *First Aid Kits* should be available in predefined locations within the workspace.

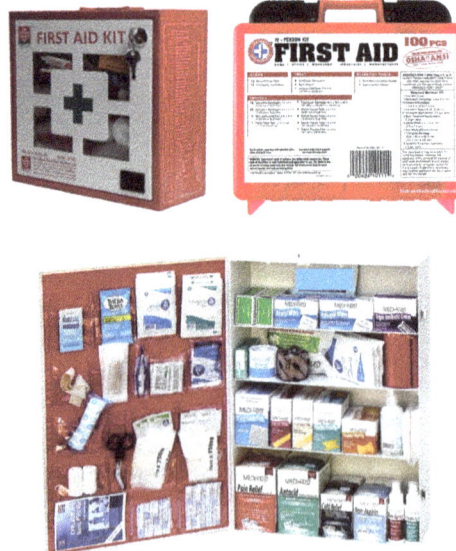

Fig. 1.49 –Industrial First Aid Kits in the Workspace

65

65

1.10.4- Fire Fighting Equipment

Must be readily available and periodically inspected.

Fire Blanket must be easily accessible.

Chemicals, Solvents, Fuels, etc.

Fig. 1.50 – Fire Fighting Equipment in the Workspace

66

1.10.5- Secured Hazardous Areas

OSHA Standard 1910.307

Fences around dangerous work zones.
(Cutting, Pressing, Crushing, Shearing, and Flying Objects).
Some work zones accessible only by authorized persons.

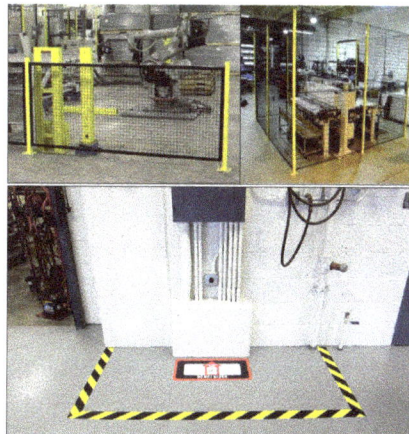

Fig. 1.51 – Secured Hazardous Area

67

1.10.6- Safety and Job Performance Posters

- Safety poster and signs to remind the employees about their safety.

- Job performance posters to increase the awareness of hydraulic machine operators about their job duties.

- Both types of posters should be:
  - Demonstrative.
  - Convey the message to people of different backgrounds.
  - located in visible places.

Fig. 1.52- Informative Posters and Signs

68

68

1.11- BP-Safety-05: Safe Startup of Hydraulic Systems

After safe design, safety of operator, safety of work environment, and safety of workspace,

the machine is ready for commissioning.

Many systems were found to work improperly after the first startup or during the testing period.

Incorrect commissioning during startup can result in damaging hydraulic components.

Usually the consequences of improper system commissioning are seen when its too late and the component may need to be replaced!

69

69

BP-Safety-05

1. Emergency Stop: Locate the emergency stop button.
2. Local Safety Instructions: Review
3. Startup Instructions: Review start up instructions if found.
4. Communication: Arrange for proper communication with people in charge.

Fig. 1.53 – Safe Commissioning of Hydraulic System (Courtesy of the International Hydraulic Safety Authority)

70

5. Oil Level: Check oil level in the reservoir and make sure it is above the minimum level.

6. Priming: Prime pumps, motors, and cylinders as per the manufacturer' instructions.

7. Hydraulic Lines: Make sure they are properly tightened.

8. Leakage Gutters: Use leakage collection gutters or curbs around the machines so that if external leakage occurs, it shall not be hazardous.

9. Electrical Lines: Make sure they are securely connected and are separated from hydraulic lines.

10. Controls: Make sure all controls and valves are in neutral.

71

11. **Pump Direction of Rotation:** Make sure the pump direction of rotation is correct.

12. **Electric Motors:** If the prime mover is an electric motor, make sure it is wired to run in the correct direction. Start and stop to confirm that the motor runs in the right direction.

13. **Safe Position:** Take safe position to avoid unexpected actuators or machine movement. All unnecessary personnel must stay away of the system.

14. **Mechanical Safety Locks:** If the machine should move, remove all the safety interlocks.

15. **Easy Start:** Test the system for at least 10 minutes under conditions of no-load, low pressure, and low RPM.

72

72

16. **Variable Pumps:** For a variable pump or motor that receives external pilot pressure, carefully and safely untighten the fitting of the **pilot line** at the components to **remove air**.

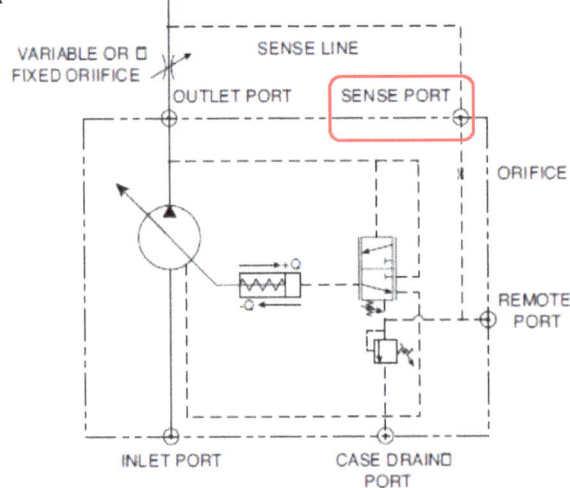

Fig. 1.54 – Pilot Pressure for a Load Sense Variable Pump (Courtesy of Parker) 73

73

17. Closed Circuits: On closed circuits (hydrostatic transmissions), monitor the charge pressure. If the charge pressure specified by the manufacturer is not established within 20 to 30 seconds, shut down the prime mover and investigate the problem. Charge pressure is typically 110-360 PSI (8-25 bar). Losing charge pressure will result in pump cavitation.

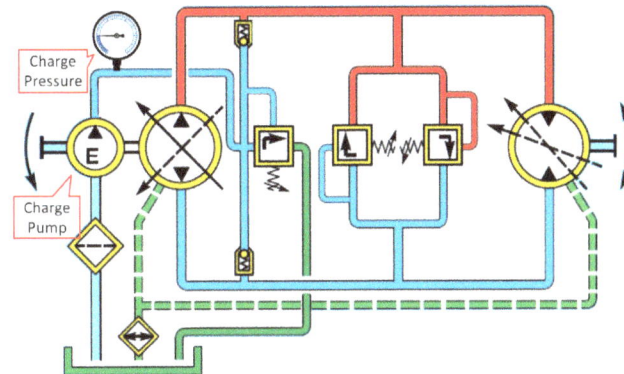

Fig. 1.55 – Charge Pressure in Hydrostatic Transmission
(Courtesy of Bosch Rexroth)

74

74

18. Easy Actuation: Stroke cylinders slowly and run hydraulic motors at low speed until all air is removed from the components and the plumbing, and the actuators move smoothly.

19. Inspection: Inspect the system for leakage, unusual noise, vibration or unusual smell.

20. Gradual Loading: Increase the load gradually until the machine runs safely under full load and maximum pressure.

21. Operating Temperature: Observe the rate of temperature increase in the system.

75

75

1.12- BP-Safety-06: Safe Operation of Hydraulic Systems

After safe start up a hydraulic-driven machine, it must continue to operate properly and safely. *BP-Safety-06* presents guidelines for safe operation of hydraulic systems.

PB-Safety 06:

1. Locate Emergency Shut Off.

2. DO NO Operate Leaking Machine.

3. Technical and Safety Training.

4. Continuous Condition Monitoring.

5. Continuous Contamination Control.

6. Periodic Inspection and Maintenance.

7. Reporting.

The following subsections provide detailed interpretation of the action items listed in BP-Safety-06.

76

76

1.12.1- Locate Emergency Shut Off

If the machine is electrically driven, the machine operator must be familiar with the locations of
- the machine emergency stop.
- the main circuit breaker for the workplace.

Machine Emergency Stop   Main Emergency Shut Off

Fig. 1.56 - Emergency Shut Off

77

77

1.12.2- DO NOT Operate a Leaking Machine

Keep all hydraulically operated equipment and surrounding areas clean and free of fluid residue and combustible materials. DO NOT operate a leaking hydraulic system. Serious injury may result.

Fig. 1.57 - Leaking Hydraulic System

78

78

1.12.3- Technical and Safety Training

Lack of understanding of "How it Works?" is a primary reason for unsafe operation of a hydraulic system.

Technical Hand-On Training:
- Basic training about the construction and operation of hydraulic systems.
- Steps necessary to perform the job.
- Hazards associated with the job.

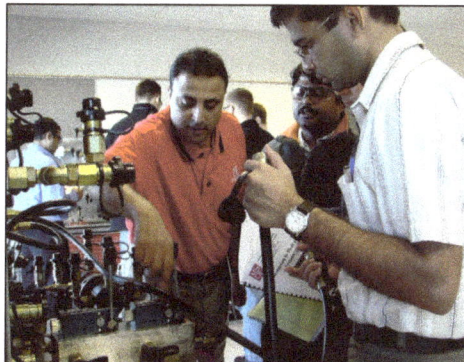

Fig. 1.58 – Hands-On Technical Training

79

79

Safety Training:
- BP-Safety-02: Safety of Hydraulic System Operators.
- BP-Safety-04: Safety of Hydraulic System Workspace.
- Emergency Plan: predefined duties for each individual within the workspace in response to various emergency situations such as fire, explosion, injuries, etc.

International Hydraulic Safety Authority (IHSA)
http://www.hsac.ca/
provides safety training focused on specific hydraulic systems applications. Such as safety of hydraulic systems in steel mills, mining, etc.

1.12.4- Continuous Condition Monitoring

1.12.5- Continuous Contamination Control

1.12.6- Periodic Inspection and Maintenance

1.12.7- Reporting

80

80

1.13- BP-Safety-07: Safe Servicing of Hydraulic System

PB-Safety 07:
1. Understand the Job Tasks and Tools.
2. Understand the Job Hazard.
3. Review Safety of the Operator and Workspace.
4. Follow Lockout Procedure
5. Secure Overrunning Loads and Moving Parts.
6. Release Stored Energy.
7. Depressurize the System.
8. Wait Until the Machine Cooled Down.
9. Prepare Service Location.
10. Prepare Service Spare Parts.
11. Prepare Service Utilities.
12. Be aware of the Common Mistakes during Hydraulic System Maintenance.
13. Avoid Oil Spillage.
14. Careful Welding.
15. Proper Cleaning and Painting.
16. Proper Storage and Transportation.

81

81

1.13.1- Understand Job Tasks and Tools

It is extremely important to understand how to perform your job safely. Therefore, as best practices, maintenance work force should:

- **Review History.**

- **Review Service Instructions.**

- **Review Best Practices for Disassembling and Assembling.**

- **Review Circuit Diagram.**

- **Review Proper Tools.**

- **Ask Questions.**

82

82

Example 1: Proper lifting of hydraulic components reduces the risk of injuries.

Risk of injury!

During transport with a lifting device, the axial piston unit can fall out of the lifting strap and cause injuries.

- ▶ Hold the axial piston unit with your hands to prevent it from falling out of the lifting strap.

- ▶ Use the widest possible lifting strap.

WARNING!

⚠

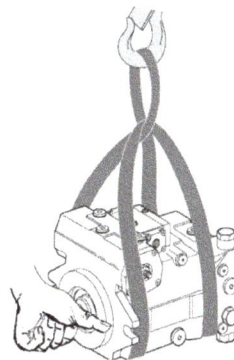

**Fig. 1.59 – Proper Transportation of Heavy Hydraulic Components
(Courtesy of Bosch Rexroth)**

83

83

Example 2: Use *Torque Wrench* to tighten bolts to the specified value.

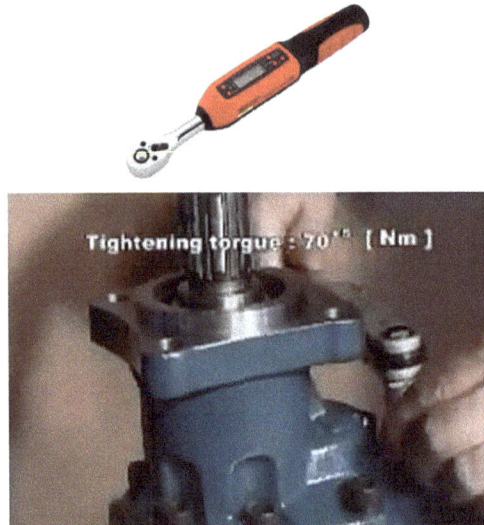

Tightening torque : 70⁴⁵ [Nm]

Fig. 1.60 – Using Torque Wrenches to Assemble Hydraulic Components

84

84

Example 3: Use the right support, otherwise load may fall.

Identify which is the right support stand and why

The stand on the left does not have an identification plate
which lists a)manufacturer, b) maximum load, c) machine &
support locaion, d) part number e) keyed to fit a specific
location

**Fig. 1.61 – Use the Right Support
(Courtesy of the International Hydraulic Safety Authority)**

85

85

Example 4: Use the right hoisting device. Otherwise, unexpected load movement could occur.

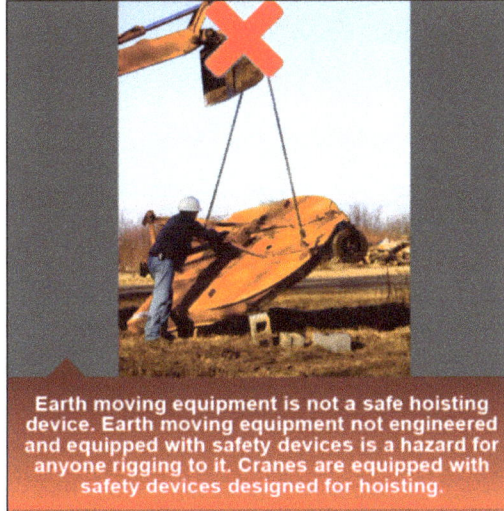

Earth moving equipment is not a safe hoisting device. Earth moving equipment not engineered and equipped with safety devices is a hazard for anyone rigging to it. Cranes are equipped with safety devices designed for hoisting.

Fig. 1.62 – Use the Right Hoisting Device
(Courtesy of the International Hydraulic Safety Authority)

86

86

Reported Case History

Action:
A maintenance technician installing a new cylinder on a production machine.

Result:
- When the valve shifted to test the cylinder, the cylinder did not move for few seconds, then accelerated and damaged the cylinder head.
- The technician suffered a non-injury accident.

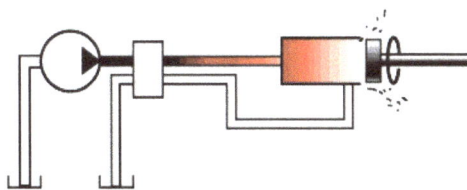

Air in transmission lines can cause an extreme safety hazard!

Fig. 1.63 – Improper Cylinder Air Bleeding
(Courtesy of Fluid Power Training. Inc.)

Reason:
- Improper bleeding of air out of the cylinder.
- Pump flow pushes the air in the transmission lines ahead of the oil compressing it.
- When pressure overcame the initial resistance to move the cylinder rod, the compressed air expanded, causing the cylinder rod to accelerate at high velocity.

87

87

Reported Case History

Action:
A maintenance technician, removed a hose from the cylinder to inspect piston seal leakage.

Result:
No injuries, the fire caused $3.5 million in damages.

Reason:
The damaged seal allowed oil to spray out through the open port at high velocity. The "atomized" oil ignited when it came into contact with the gas heater that was mounted above the machine.

FPSI™ © 2002

**Fig. 1.64 – Improper Cylinder Leakage Inspection
(Courtesy of Fluid Power Training. Inc.)**

88

88

Reported Case History

Action: A millwright is testing case drain leakage of a hydraulic motor on a conveyor system.

Result: The technician suffered an eye injury, minor burns, bruises, and abrasions as a result of an accident.

**Fig. 1.65 – Improper Motor Leakage Inspection
(Courtesy of Fluid Power Training. Inc.)**

Reason: The technician improperly inspect the case drain leakage of the motor. He disconnected the case drain line from the port at the motor housing and held it in a receptacle. After running the system for few moments, without warning, the case drain oil surges and began to spray out in his face with such intensity that it jerked the hose violently.

89

89

1.13.2- Understand Job Hazards

knowing the hazards of the job → Reduces the accidents.

❑ Supervisor should:
- Explain the hazards, how to avoid them, and how to respond.
- Point out to parts that could move during your servicing of the system.

General Job Hazards:
General job hazards when servicing a hydraulic system:

Example 1: Hazards from Electrical Cords
- Frayed cords or damaged connectors are dangerous. Must be replaced.

Fig. 1.66 – Hazards from Electrical Cords

90

90

Example 2: Hazards from Control Panels

Unexpected loss of control due to rusty or corroded control panel could result in serious injuries, particularly if the control is safety-related.

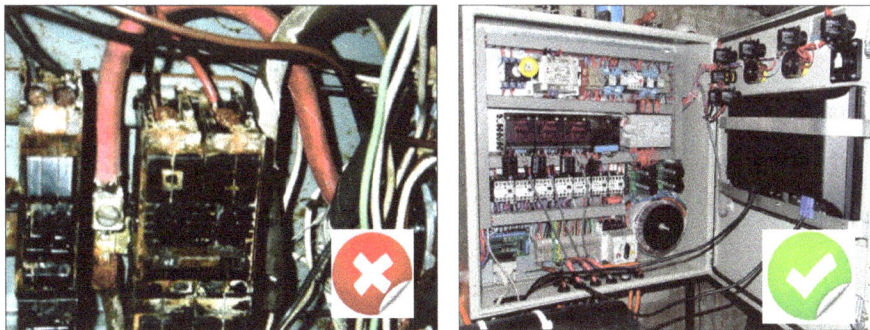

Fig. 1.67 – Hazards from Electrical Control Panels

91

91

Example 3: Hazards from Hand Pumps

Many of the workers are not aware that hand pumps can generate extreme high pressure. Improper use of such pumps could result in serious injuries.

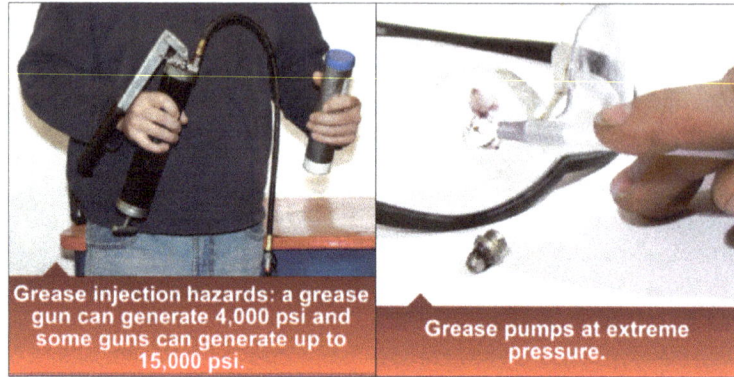

Grease injection hazards: a grease gun can generate 4,000 psi and some guns can generate up to 15,000 psi.

Grease pumps at extreme pressure.

**Fig. 1.68 – Hand Pump Generates Extreme Pressure
(Courtesy of the International Hydraulic Safety Authority)**

92

92

Example 4: Hazards from Unfastened Hoses

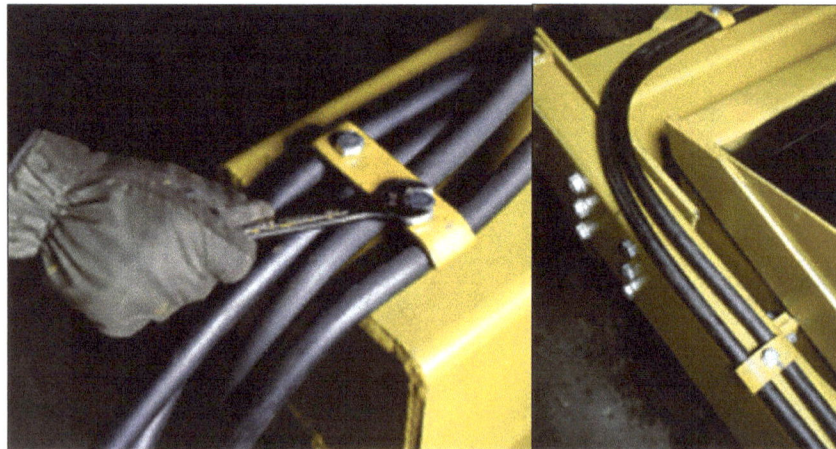

Fig. 1.69 – Hydraulic Hose Clamps Prevents Whipping if Hose Fails

93

93

1.13.3- Review Safety of the Operator and Workspace

- DO NOT service a hydraulic-driven machine where safety regulations for the operator and the workspace are not established.

- Before servicing a hydraulic driven machine, review available safety regulations for servicing personnel and the workspace.

- If no regulations found, review the guidelines presented by BP-Safety-02/03/04/05/06.

Fig. 1.70 – Safety of Operator and Workspace

94

94

1.13.4- Follow Lockout Procedures

Lockout Procedure.

If no instructions available, follow the shown below instructions:

- Lockout tags.
- Turn off prime movers.
- Disconnect from main electric power.
- Disconnect from compressed air sources.

V201 (0.5 min)

Fig. 1.71 – Lockout Tags

95

95

Reported Case History

Action:

A millwright was standing on a stepladder tightening a leaking hydraulic connector in a steel hydraulic transmission line that was affixed to a wall 12 feet above the floor.

Result:

- The connector unexpectedly failed.
- High-pressure hydraulic oil sprayed from the broken connector, striking him in the face and chest.
- He lost his grip on the ladder, and fell to the concrete floor below.
- He died as a result of the injuries he sustained from the fall.

Reason:

Power unit left ON during maintenance.

Fig. 1.72 – Hazard of Maintaining a Running Hydraulic System (Courtesy of Fluid Power Training. Inc.)

96

96

1.13.5- Secure Overrunning Loads

- **Example 1:** A telescopic hydraulic cylinder of a hydraulic-driven elevator must lowered to the last stage before servicing the hydraulic system.

- **Example 2:** A loader bucket must be lowered to the ground before servicing the loader.

- **Example 3:** A conveyor should be clamped before servicing the hydraulic drive system.

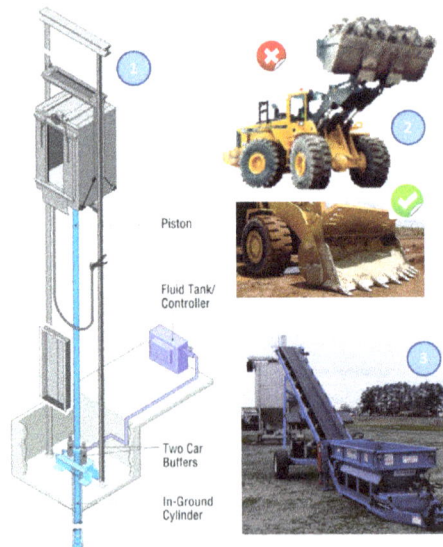

Fig. 1.73 – Hazard of Maintaining a Running Hydraulic System

97

97

1.13.6- Release Stored Energy

Stored Energy is a HIDDEN ENEMY !!

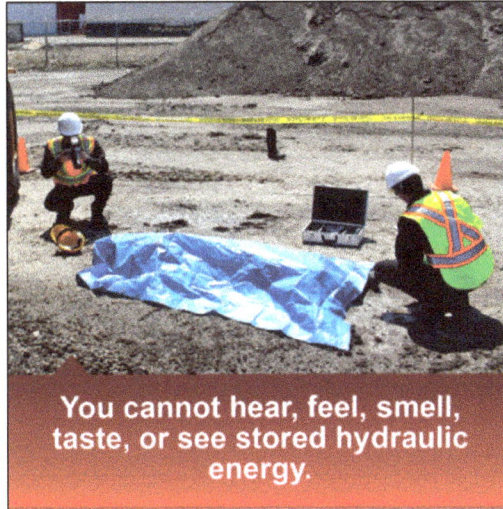

You cannot hear, feel, smell, taste, or see stored hydraulic energy.

Fig. 1.74 – Hazard from Stored Energy
(Courtesy of the International Hydraulic Safety Authority)

98

98

Accidental or unprofessional releasing of such energy may result in move of actuators and serious injuries.

- **Example 1: mechanical** stored energy in spring-return single-acting cylinders.
- **Example 2: fluid** stored energy in accumulators.
- **Example 3: gas** stored energy in accumulators.

Fig. 1.75 – An Example of Unprofessional Release of Stored Energy
(Courtesy of Fluid Power Training. Inc.)

99

99

Reported Case History

Action:
disassembling piston-type, gas-charged accumulator.

Result:
A maintenance helper lost his hand as a result of an accident he suffered.

Reason:
Energy stored in the accumulator released in very short time not in accordance with OSHA regulations.

Fig. 1.76 – Hazard from Unprofessional Gas Discharge from an Accumulator (Courtesy of Fluid Power Training. Inc.)

100

100

1.13.7- Depressurize the System

After lowering all suspended loads and releasing all stored energies, there may still be some residual pressure present inside some parts in the system.

Always verify zero hydraulic energy as part of the lockout procedure.

Fig. 1.77 – Depressurize the System before Service (Courtesy of the International Hydraulic Safety Authority)

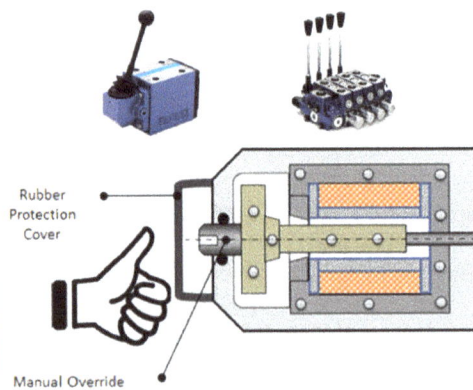

Rubber Protection Cover

Manual Override

Fig. 1.78 – Move Directional Valves to Release Trapped Pressure

V200 (1 min)

101

101

1.13.8- Wait Until the Machine Cools Down

Dimensions and characteristics change when hot components are assembled

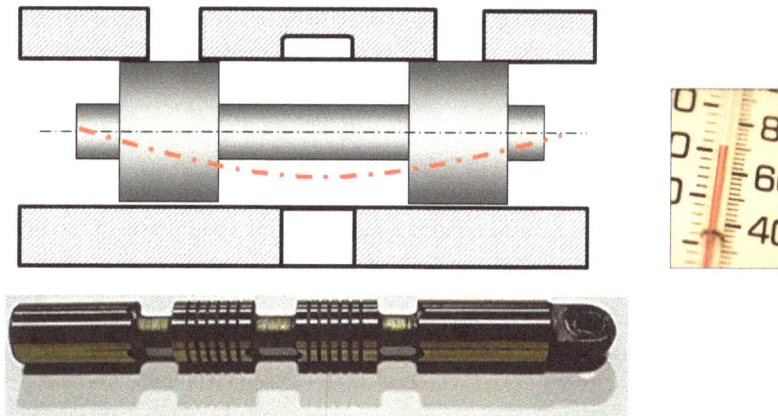

Fig. 1.79 – Dimensional Change when Dissembling a Hot Component

102

102

1.13.9 - Prepare Service Location

Servicing area must be clean, dray and organized

Fig. 1.80 – Organized Service Location Helps to avoid Risks of Injuries and Contamination

V156 (0.5 min)

103

103

1.13.10- Prepare Service Spare Parts

- Manufacturer's instructions.

- Part numbers match.

- Genuine spare parts.

- Certified or trusted sources.

- Using non-original spare parts may result in voiding the warranty.

- Some part are non-reusable.

Genuine Spare parts

Cutting (Sealing) Metallic Rings   Rubber O-Rings

Fig. 1.81 – Importance of using Genuine Spare Parts in Servicing Hydraulic Systems

104

104

1.13.11- Prepare Service Utilities

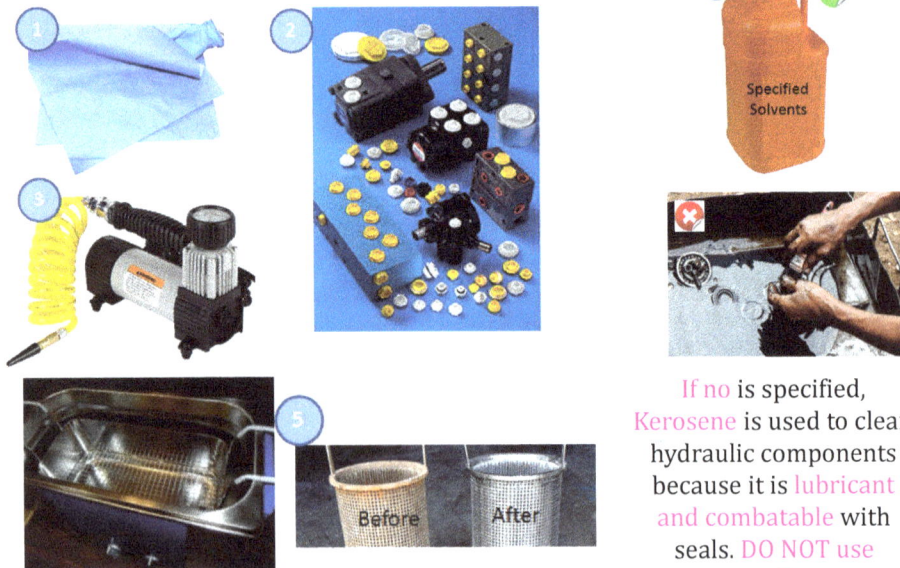

Specified Solvents

If no is specified, Kerosene is used to clean hydraulic components because it is lubricant and combatable with seals. DO NOT use gasoline.

Before   After

Fig. 1.82 – Examples of Maintenance Utilities

105

105

1.13.12- Be aware of Common Mistakes during Hydraulic System Maintenance

Avoid common mistakes during system maintenance. Such common mistakes will be reported in the next volume of this textbook series.

106

106

1.13.13- Avoid Oil Spillage

Sources of spilled oil:
- Leaking hydraulic systems.
- Disassembled hydraulic components.

Result of spilled oil:
- Slippage and falling down.
- Spilled oil may cause fire.
- Money wasting.

Best Practices to avoid risks from spilled oil:
- Prepare for collecting oil from the disassembled components.
- Oil waste container.
- Dispose of used oil and system filters as required by federal, state, or local regulations.
- Dry hands: no slippage and transport contamination to components.
- When the service work is completed, thoroughly clean any spilled oil from the equipment.

107

107

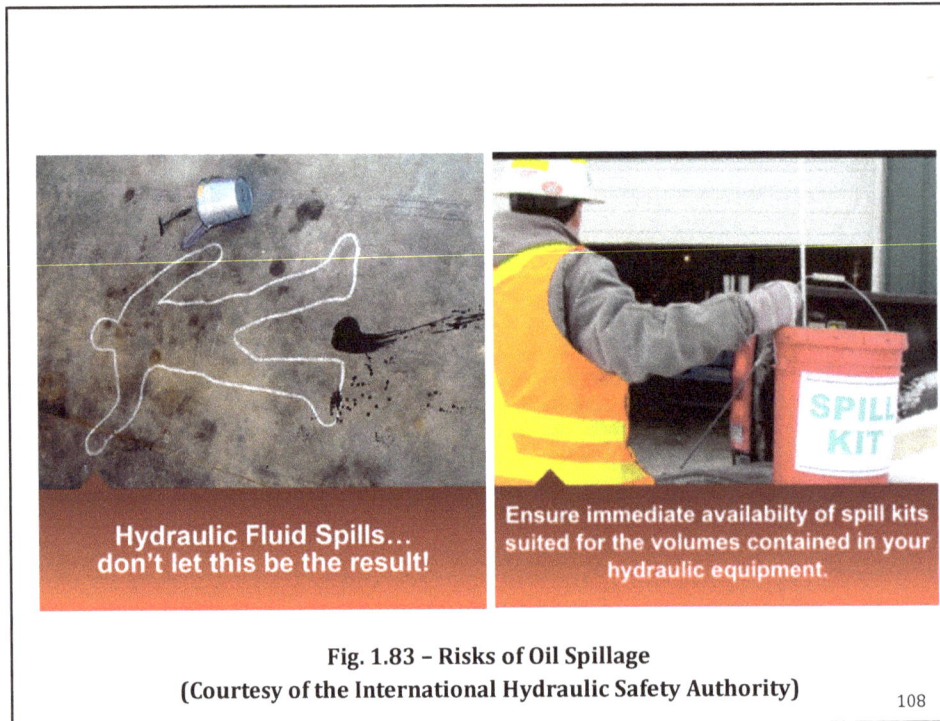

Fig. 1.83 – Risks of Oil Spillage
(Courtesy of the International Hydraulic Safety Authority)

108

108

1.13.14- Careful Welding

Do not cut or weld in any area where hydraulic fluids are used until the area is free of all oil deposits and the system is shut down and depressurized.

109

109

1.13.15- Proper Cleaning and Painting

Fig. 1.84 – Protect Sensitive Surfaces during Painting

110

110

1.13.16- Proper Storage and Transportation

Improper packaging, transportation, and storage may result in damaging hydraulic components. The next volume of this textbook series provide best practices to cover this subject for the various hydraulic components. Generally speaking as best practices:

Packaging: All parts of the hydraulic system shall be packaged for transportation in a manner that preserves their identification and protects them from damage, distortion, contamination and corrosion.

Transportation: Follow the given instructions. The following are just two examples:
- Accumulators should be discharged from both gas and oil before shipping.
- Hydraulic power units should not be filled with oil during shipping.

Storage: Follow the given instructions for each hydraulic components.

111

111

1.14- BP-Safety-08-Oil Injection Avoidance and Treatment

How Oil Injection Happens?

Fine streams of escaping pressurized fluid can penetrate the skin and thus enter the human body causing serious injuries, loss of organs, or even death.

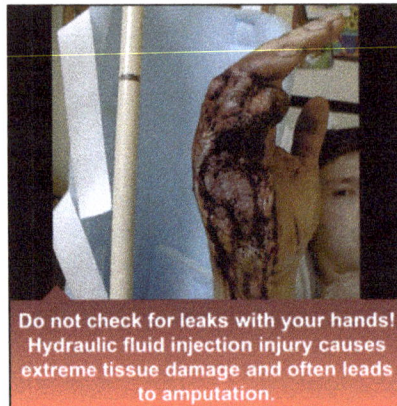

V272 (1 min)

Do not check for leaks with your hands! Hydraulic fluid injection injury causes extreme tissue damage and often leads to amputation.

Fig. 1.85 – Risks of Oil Injection
(Courtesy of the International Hydraulic Safety Authority)

112

112

Challenges of Oil Injections:

- According to the "Occupational Injuries Handbook," oil can penetrate the skin at pressures as low as 100 PSI (7bar).

- The symptoms of oil injection sometimes don't appear until the injected area becomes in critical condition.

- Fluid injected into the skin must be surgically removed within hours or gangrene may occur resulting in amputation of the affected area.

- Some medical care providers are unfamiliar with fluid injection injuries.

113

113

BP-Safety-08-Oil Injection Avoidance and Treatment

Best practices to avoid oil injection:
- Make the people aware of the oil injection risks. V262 (1 min)
- Never use your hand or fingers to search for leak.
- Wear protective gloves when servicing hydraulic transmission lines.

Best practices if fluid injection occur:
- Report the case immediately to your supervisor
- DO NOT allow the injured person to drive himself to the medical facility.
- DO NOT give food or drink for the injure person.
- DO NOT treat injections as a simple cut!
- DO NOT delay treatment.
- See a specialist doctor immediately.
- Prepare the case information as shown in the safety card provided by the International Fluid Power Safety (IFPS).

114

114

Fig. 1.86 – Safety Focus Card
(Courtesy of the International Fluid Power Society)

115

115

CRITICAL INFORMATION

You must get at least 5 critical pieces of information for doctors.

WHAT TYPE OF FLUID?
Grease, oils and hydraulic fluid often don't cause any initial reaction, paint and paint thinners/solvents dissolve fat and will cause early, intense inflammatory reactions.

WHAT IS THE AMOUNT OF FLUID INJECTED?
The more fluid injected, the less room for blood circulation.

WHAT WAS THE PRESSURE OF FLUID INJECTED?
Pressure of the fluid may help determine the amount injected.

WHAT IS THE DEGREE OF SPREAD OF INJECTED MATERIAL? Location of the injury will help determine the degree of spread. Cases have been reported where fluid injected into the hand has been recovered as far away as the elbow.

HOW MUCH TIME LAPSED BETWEEN INJECTION AND TREATMENT? This is an extremely important factor determining the outcome. The sooner that surgery is carried out post-injection, the less long-term disability will result.

116

116

V132 (12 minutes)

Reported Case History

Action:
A technician used his hand to check leakage from a hose spraying oil.

Result:
2 fingers lost.

Reason:
High pressure oil injection not treated immediately.

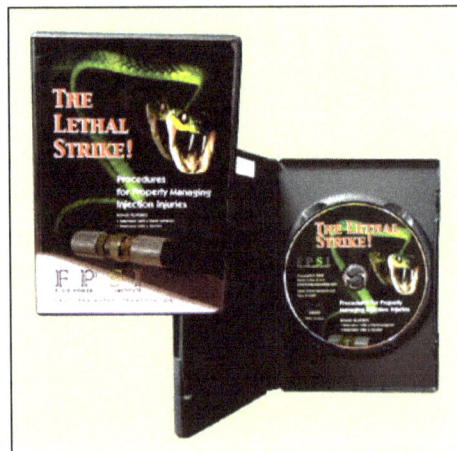

Fig. 1.87 – Oil Injection Training Video
(Courtesy of Fluid Power Training. Inc.)

117

117

1.15- BP-Safety-09: Safe usage of Hydraulic Powered Tools

❏ Hydraulic powered hand tools such as:

- Lifting cylinders.
- Bolting tools.
- Pullers, work.
- Holding tools.
- Custom tools.

❏ Are widely used in various applications such as:

- Civil engineering,
- Mining,
- Oil and gas,
- Power generation
- Shipbuilding, etc.

V266 (1 min)

**Fig. 1.88 – Hydraulic Powered Tools
(Courtesy of ENERPAC)**

❏ Improper use of hydraulic powered tools under high pressure may results in serious injuries.

118

118

BP-Safety09:

❏ BP-Safety09-A: Safe Operation of Cylinders in Hydraulic Powered Tools.

❏ BP-Safety09-B: Safe Operation of Hoses in Hydraulic Powered Tools.

❏ BP-Safety09-C: Safe Operation of Pumps in Hydraulic Powered Tools.

119

119

1.15.1-BP-Safety-09-A: Safe Operation of Cylinders

Cylinders in hydraulic powered tools
vary widely in terms of sizes and strokes

- **Cylinder Inspection:** Inspect the cylinder of the tool for signs of wear, corrosion or damage.

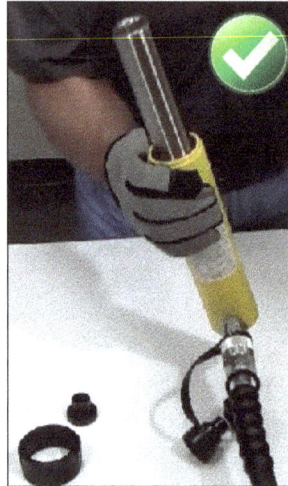

**Fig. 1.89 - Cylinder Inspection
(Courtesy of ENERPAC)**

120

120

- **Modified Cylinders:** Never use cylinders that have been modified.

**Fig. 1.90 - Modified Cylinders
(Courtesy of ENERPAC)**

121

121

- **Air Bleeding:** Bleed air from the cylinder before first use as per the instruction manual.

Fig. 1.91 - Cylinder Air Bleeding (Courtesy of ENERPAC)

122

122

- **Saddle Usage:** Properly use a saddle to ensure even load distribution on the cylinder plunger. Without a saddle, the cylinder rod may be permanently damaged.

Fig. 1.92 - Importance of using a Saddle (Courtesy of ENERPAC)

123

123

- **Saddle Placement:** The entire saddle must be in full contact with the load. Placing only portion of the saddle under the load puts regular stress on the plunger, which can damage equipment.

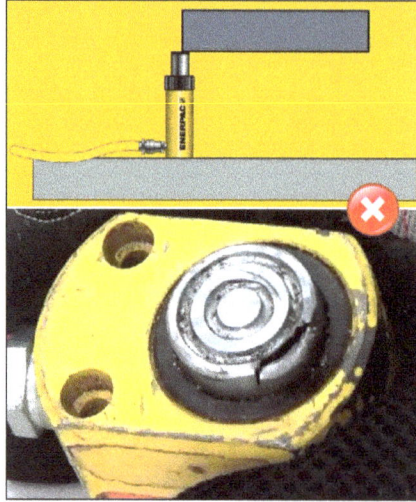

Fig. 1.93 - Hazard from Improper Saddle Placement (Courtesy of ENERPAC)

124

124

- **Cylinder Placement:** Make sure the surface under the cylinder is flat as possible and use the cylinder support bases to offer more stability.

Fig. 1.94 - Proper Cylinder Placement (Courtesy of ENERPAC)

125

125

- **Side Loading:** Avoid side loading. If the cylinder rod is bent, stop pumping, retract the load, and determine what changes are required to complete the task safely.

Fig. 1.95 - Effect of Side Loading (Courtesy of ENERPAC)

126

126

- **Maximum Load:** Do not exceed 80% of the manufacturer stroke and load levels.

Fig. 1.96 - Rule for Maximum Load (Courtesy of ENERPAC)

127

127

- **Pump Handle:** Remove the handles when it is not being used to avoid creating tripping hazard.

Fig. 1.97 - Remove Handle when if not needed (Courtesy of ENERPAC)

128

128

- **Body Protection:** Never put your body under the load without making sure that the load in mechanically secured.

Fig. 1.98 - Body Protection (Courtesy of ENERPAC)

V267-Operation Safety of Cylinders in Hydraulic Tools (2.5 min)

129

129

1.15.2- BP-Safety-09-Powered Tools-B: Safe Operation of Hoses

Hoses in hydraulic powered tools are the most source of unsafe incidents in hydraulic powered tools.

- **Hose Inspection:**
  o Inspect the hose for cuts, cracks, abrasion, or any signs of damage.

  o It is recommended not to use 6 years old hoses no matter its condition.

Fig. 1.99 - Hose Inspection (Courtesy of ENERPAC)

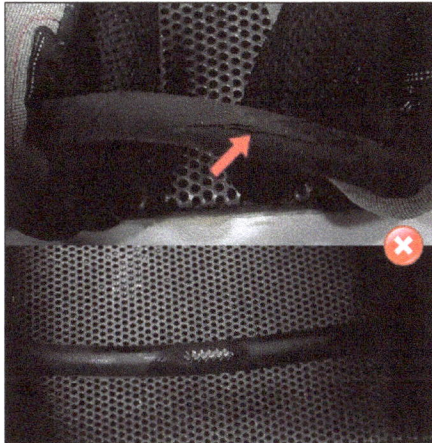

130

130

- **Tool Handling:** Never lift, carry, or drag equipment from a hose.

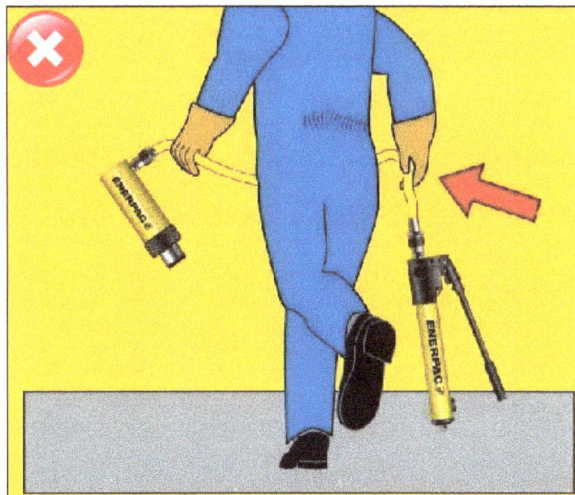

Fig. 1.100 - Improper Tool Handling (Courtesy of ENERPAC)

131

131

- **Care of Hoses:** Avoid dropping objects on hoses or driving equipment over a hose.

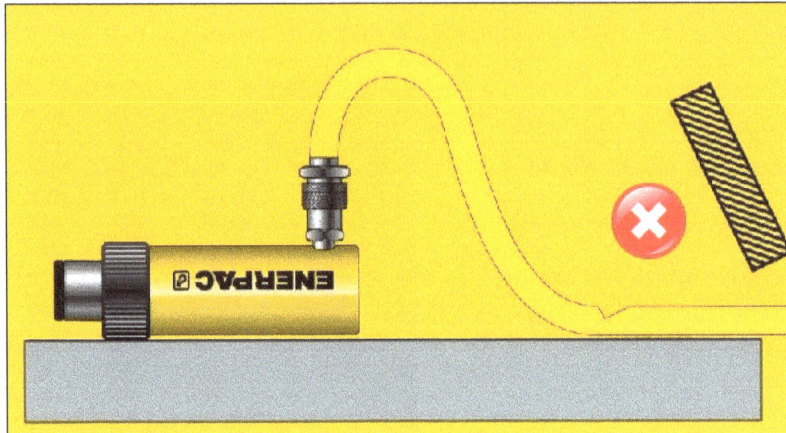

Fig. 1.101 - Improper Care of Hoses (Courtesy of ENERPAC)

132

132

- **Hose Placement**: Keep hoses away from areas enclosed by the lifted load.

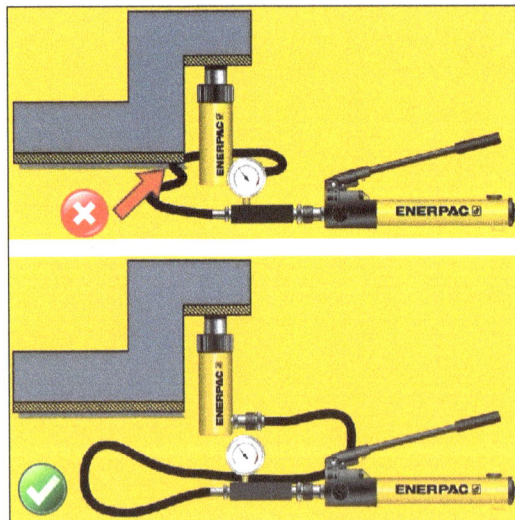

Fig. 1.102 - Hose Placement (Courtesy of ENERPAC)

133

133

- **Minimum Bend Radius:** Avoid short bends or kinking hoses. DO NOT exceed the minimum bend radius specified by the manufacturer.

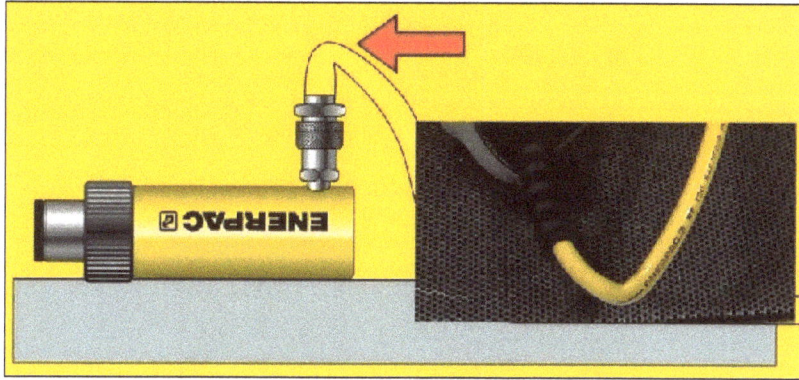

Fig. 1.103 - DO NOT Exceed Minimum Bend Radius of the Hose (Courtesy of ENERPAC)

134

134

- **Hose Rupture:** If a hydraulic hose is ruptures, immediately release system pressure and retract the load.

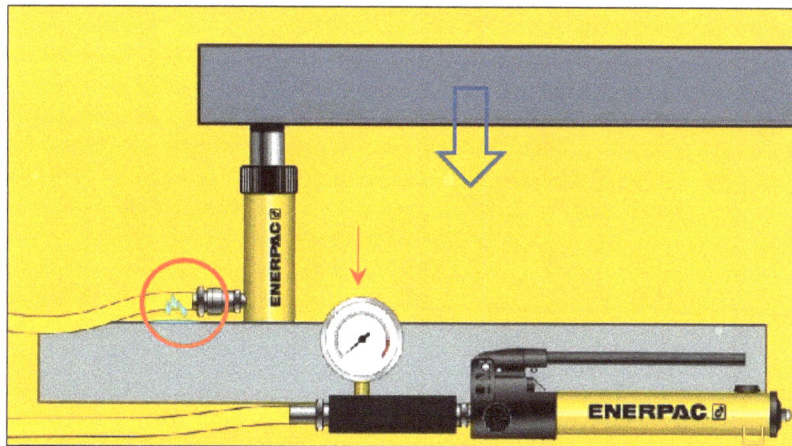

Fig. 1.104 - What to do if a Hose Ruptures (Courtesy of ENERPAC)

135

135

- **Hose Detachment:** Never detach a hose or a coupling while the system still under pressure.

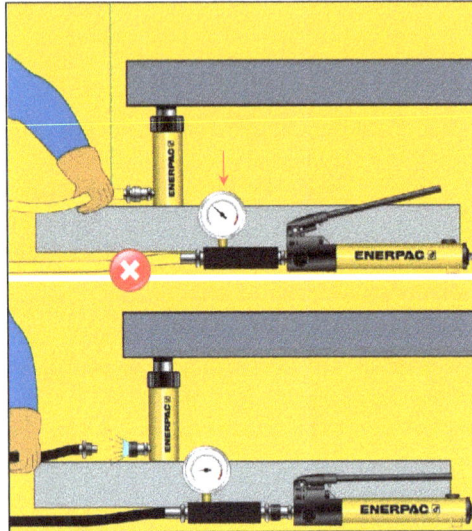

Fig. 1.105 - Improper Hose Detaching (Courtesy of ENERPAC)

136

136

- **Inspect Hose Couplings:** Before attaching couplings, check both parts to make sure there are no signs of defects or damage.

Fig. 1.106 - Inspect Coupling before Connecting (Courtesy of ENERPAC)

137

137

- **Dust Caps:** Use dust caps when the couplings are not used.

**Fig. 1.107 - Use Protective Caps When Couplings are not Connected
(Courtesy of ENERPAC)**

138

- **Improper Hose Coupling:** Never use low pressure couplings or fittings that are not compatible with high-pressure operation.

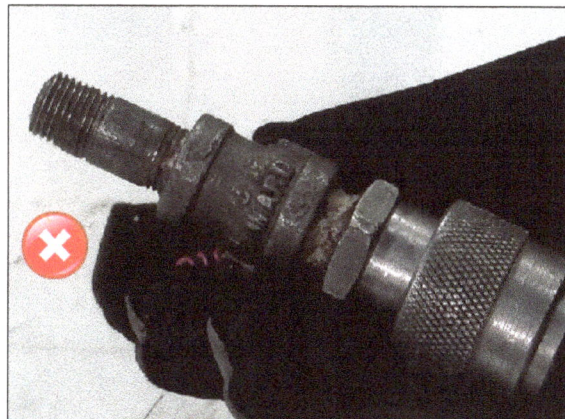

**Fig. 1.108 - Use Fittings and Couplings Rated for System Operative Pressure
(Courtesy of ENERPAC)**

V268-Operation Safety of Hoses in Hydraulic Tools (3 min)

139

1.15.3- BP-Safety-09-Powered Tools-C: Safe Operation of Pumps in Hydraulic Powered Tools

Pumps in hydraulic powered tools could be hand-operated, electrical powered, battery powered, or air powered.

- **Pump Priming:**
  o If the pump is connected to a cylinder, make sure the cylinder is fully retracted before adding oil.

Fig. 1.109 - Retract Attached Cylinder before Filling the Pump
(Courtesy of ENERPAC)

140

140

o When using a hand pump, fill it only to recommended level. Too much oil limit performance and may result in a leak.

Fig. 1.110 - Fill the Pump to Recommended Level

(Courtesy of ENERPAC)

141

141

o   Close the release knob moderately. Too much force will damage the valve.

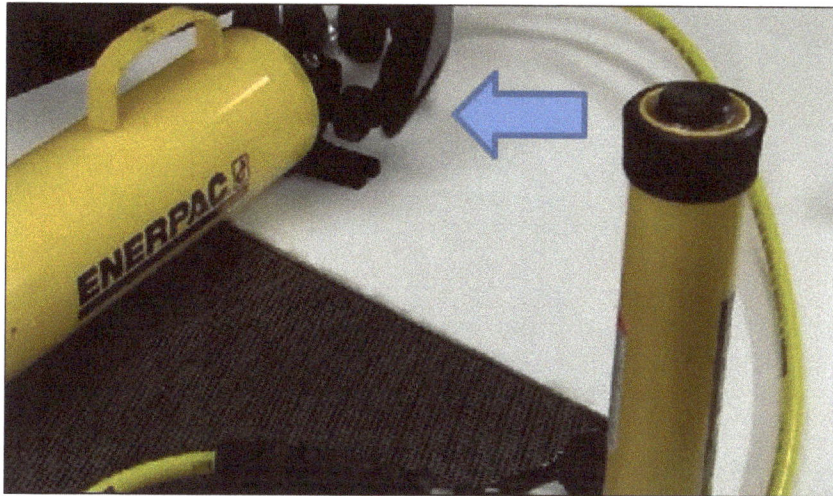

Fig. 1.111 - Close the Release Knob Moderately
(Courtesy of ENERPAC)

142

142

- **Pump Handle:** Never use an extension on the handle for more leverage. Hand pumps are designed to work efficiently with the built in handle.

Fig. 1.112 - DO NOT Use Extensions on Pump Handles
(Courtesy of ENERPAC)

143

143

- **Pressure Setting:** When force setting is critical, always use a calibrated pressure gauge. Never override the factory setting of the relief valve and do not exceed cylinder stroke length.

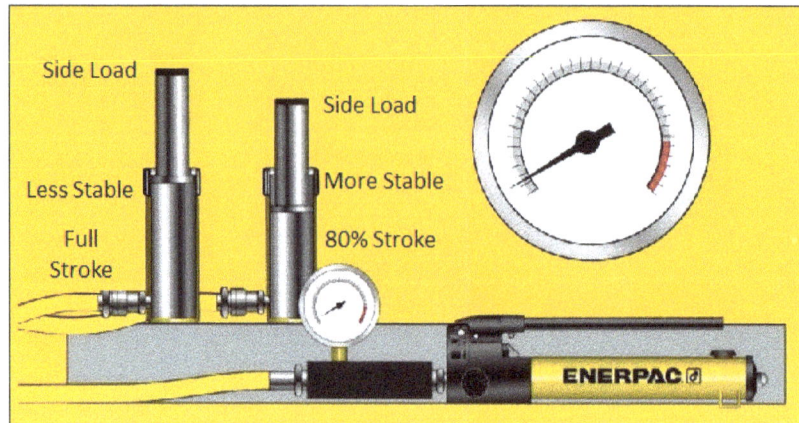

**Fig. 1.113 - DO NOT Exceed 80% of the Cylinder Stroke
(Courtesy of ENERPAC)**

144

144

- **Electrical Connection:** Make sure that the electrical cord is not damaged and the plug matches the electrical receptacle voltage and the current requirements.

Fig. 1.114 - Proper Electrical Connection (Courtesy of ENERPAC)

145

145

- **Pump Battery:** For cordless electric pumps, follow the manufacturer's guide for battery, charger, and pump operation.

Fig. 1.115 - Battery-Operated Pump (Courtesy of ENERPAC)

146

146

- **Pump Start Up:** Regardless of the type of the powered pump, never start the powered pump for first time under load. Always make sure that the directional control valve is in neutral position and it matches the application.

Fig. 1.116 - Never Start a Powered-Pump Under Load (Courtesy of ENERPAC)

V270-Operation Safety of Pumps in Hydraulic Tools (2 min)

V269-Operation Safety of Attachments in Hydraulic Tools (1 min)

147

147

1.16- BP-Safety-10: Safe Storage and Transportation of Hydraulic Systems

- When a hydraulic power unit isn't properly lifted appropriately →
- It may lose its stability →
- It can be knocked over, fall or move in an uncontrolled way →
- Risk of injuries or life loss.

BP-Safety-10:

- **Transportation of Hydraulic Power Units:**
- **Personnel Involvement:** No unauthorized persons are allowed.
- **Oil:** It shouldn't be transported with oil in, or accumulators charged.
- **Packaging:** Review federal/state packaging and shipping instruction.
- **Weight & CG:** Check the weight and location of the center of gravity.
- **Foundation:** must insures its stability.
- **Attachments Points:** Only the intended locations and attachment points should be used for securing or lifting the hydraulic power unit. Hydraulic power units must never be attached to or lifted at the mounted components (piping, hoses, manifolds, electric motors, accumulators, etc.).

148

148

- **Storage:**
- **Ambient Conditions:** Review the ambient conditions available in the product specific documentation.

- **Package Removal:** The packaging should not be removed until directly before assembling the unit. If the package has to be opened e.g., for inspection purposes, you should reseal the packaging to the condition in which it was supplied.

- **Stability:** provide for sufficient stability before removing any packing/transit materials or fixtures.

149

149

CONCLUSION

V158 (4 minutes)

150

150

Chapter 1 Reviews

1. As best practices of safety, who is responsible for applying the safety measures in a work area?
   A. Rescue crews.
   B. Workers.
   C. Supervisors.
   D. Everyone in the work environment.

2. Which of the following statements is considered **FALSE**?
   A. Hydraulic systems are less dangerous than electrical systems.
   B. Hydraulic systems safety should be considered only during system design.
   C. Hydraulic system safety is important because it increase the service life of the machine.
   D. All the above statements are false.

3. Which one of the following statements matches designing a safe hydraulic system?
   A. A hydraulic system should be designed to work no more than 8 hours/day.
   B. In case of power outage, the system should fail safely.
   C. A hydraulic system should be designed to maintain lower working temperature.
   D. Contamination control is not an important issue for hydraulic system safe operation.

4. Hydraulic components are basically sized based on?
   A. Maximum pressure.
   B. Maximum flow.
   C. Maximum oil viscosity.
   D. Maximum temperature.

5. Energy wasted in a hydraulic system converts to?
   A. More work at the hydraulic actuator.
   B. More noise at the pump.
   C. More heat added to the system.
   D. More pressure at the valves.

6. Pump cavitation is considered as a safety issue because?
   A. The overall system could fail when a pump is cavitated.
   B. Pump cavitation generates bothering sound.
   C. Pump cavitation increases the pressure downstream the pump.
   D. Pump cavitation results in sustained foam in the system.

7. To minimize the risks of fire development, fire-resistant hydraulic fluids are specified for applications such as?
   A. Construction machines.
   B. Machine tools.
   C. Agricultural machines.
   D. Applications such as mining, die casting, hot forging.

8. The shown operator is going to work on the shown hydraulic system, which of the following collection of equipment complies with the operator's safety?
   A. 1, 2, and 3.
   B. 3, 4, and 5.
   C. 6, 7, and 8.
   D. 4, 5, and 7.

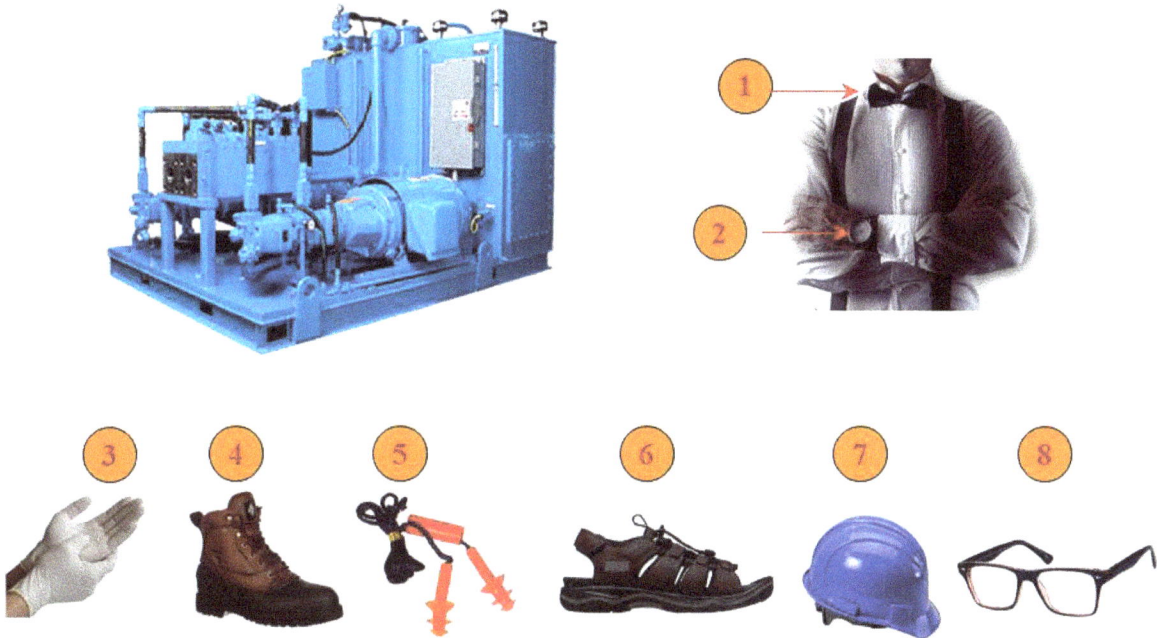

9. As per Occupational Safety and Health Administration (OSHA), the permissible daily exposure for a sound level of 90 dB is?
   A. 3 Hours.
   B. 4 Hours.
   C. 6 Hours.
   D. 8 Hours.

10. As per Occupational Safety and Health Administration (OSHA), the minimum lightening requirements in a machine shop is?
    A. 10 Watt.
    B. 10 DB.
    C. 10 FTC.
    D. 10 Joule.

Chapter 1 Assignment

Student Name: --- Student ID: ------------------

Date: --- Score: -----------------------

Assignment:

1. Which of the following statements is considered TRUE?
    A. General housekeeping has a great influence on the number of accidents and injuries in a hydraulic system workspace.
    B. Marking the floor in a workspace is not important since the workers are familiar with the place.
    C. Medical first aids and firefighting equipment must be kept in a locked cabinet that is opened only by authorized personnel.
    D. The most important thing in a workspace is to have the right tools to use.

2. Which of the following statements is considered FALSE?
    A. Before starting up a hydraulic system, oil level in the reservoir must be checked.
    B. During start up a closed hydraulic circuit, charge pressure must be checked.
    C. During start up a hydraulic circuit, if a burning smell was found, continue operating the system till the temperature raised to 110 °F.
    D. During start up a hydraulic circuit, the system should operate at no load for some time before loading it.

3. The most two important operating conditions need to be continuously monitored during the hydraulic system operations are?
    A. Contamination level and oil viscosity.
    B. Working pressure and temperature.
    C. Air in the fluid and flow rate of the pump.
    D. Sound and vibration of the system.

4. What action should be taken by a person that is exposed to oil injection into a body part?
    A. Just forget about it.
    B. Ask a friend what to do.
    C. Just ask for first aid.
    D. Immediately seek medical help and ask to see specialist.

5. The 80% rule when using hydraulic powered tool means?
    A. DO NOT load the hydraulic cylinder more than 80% of its rate.
    B. DO NOT extend the cylinder more than 80% of its stroke.
    C. DO NOT exceed 80% of the pump rated pressure.
    D. All the above.

Chapter 2
Basic Concepts of Hydraulic System Maintenance

Objectives:

This chapter covers basic rules of hydraulic system maintenance and skill set required for service workers. Impact of maintenance on system reliability and various maintenance techniques are presented. Common mistakes and reasons to void warranty are discussed. Best practices of maintaining a specific hydraulic component will be presented in the relevant chapter including guidelines for selection, replacement, installation, storage, maintenance scheduling, and standard testing.

0

0

Brief Contents:

2.1-Introduction

2.2-Hydraulic Systems Maintenance Programs

2.3-Factors that affect the Frequency of Maintenance

2.4-Hydraulic Systems Maintenance Record Keeping

2.5-Skill Set for Maintenance and Troubleshooting Team

2.6-Setting Hydraulic Systems Stages of Services

2.7-Common Mistakes during Hydraulic System Maintenance

2.8-Reasons to Void Warranty During Hydraulic System Maintenance

2.9-Standard Tests for Hydraulic Components

1

1

2.1- Introduction

2.1.1- Impact of Maintenance on Hydraulic Systems Reliability

❖ Hydraulic systems are designed to be:
- Safe
- Dry
- Efficient
- Quit
- Reliable

❖ Lack of maintenance make hydraulic systems →
- Unsafe
- Leaking
- Inefficient
- Noisy
- Less productive hydraulic systems

2

2

2.1.2- Major Causes of System Failures

Worst two enemies for a hydraulic system:
- Overheating.
- Fluid contamination.

Defending a hydraulic system against these two enemies →
improves system reliability.

2.1.3- Factors that Reduce Service Life of a Hydraulic System

Service life of hydraulic systems are drastically reduced due to one or more of
the following:

- Hydraulic fluid physical, chemical, and contamination instability.
- Poor components maintenance.
- Poor component quality.
- Abuse or improper use of the components or systems.
- Ignoring manufacturers' operational and maintenance recommendations.

3

3

2.2- Hydraulic Systems Maintenance Programs

Maintenance of hydraulic systems should be planned as follows:

- Done on predetermined bases.
- Done by qualified workers.
- Based on instructions from the system/component manufacturer.

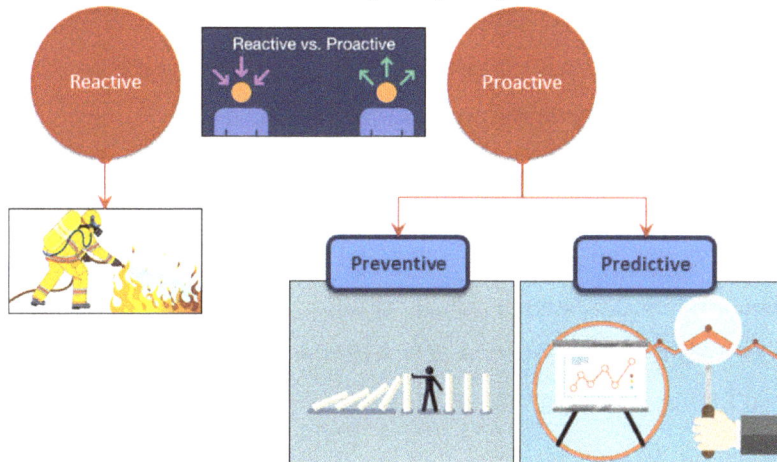

Fig. 2.1- Maintenance Methodologies

4

Reactive Maintenance:

❖ "Run-to-Failure", "Break Down", "On Demand", or "Fire Fighting".

❖ Hydraulic components are repaired as they fail →

❖ Hydraulic system can possibly fail and shutdown at any time.

Cost of this method →

- Production downtime Cost.
- Parts and material inventory.
- Labor cost.
- Costs of failure and other liabilities.

Video154 (3m.)

5

Preventive Maintenance:
Selected maintenance tasks performed routinely to avoid sudden failures:

- **Hydraulic Fluid:** Check oil level.

- **Filters Breathers, and Strainers:** Check and clean clogging.

- **Hydraulic Reservoir:** Check and clean sludge, bacteria, etc.

- **Operating Pressure:** Check and record.   Video155 (3m.)

- **Operating Temperature:** Check and record.

- **Plumbing:** Check tightness (not to the point of distortion), and leakage.

- **Outside Surfaces**: Check and keep it clean.

- **Pumps**: Check and record pump flow and suction pressure.

- **Actuators:** Check (external leakage, seals/wipers, and load connections).

- **Valves:** Check spool shifting and electrical connections for EH valves.

- **Heat Exchangers:** Check for effectiveness.

- **Noise/Vibration/Odor:** observe if any of these conditions are found.

6

6

Predictive (Smart) Maintenance:
- ❖ Routine inspections and tests to detect *root-causes* that may lead to future failures.

- ❖ Examples:
- Hydraulic fluid contamination analysis (checking cleanliness level).
- Hydraulic fluid content analysis (contents of Copper, Iron, Silicon, H_2O).
- Hydraulic fluid conditions (viscosity, additives, and oxidation).
- Wear and failure analysis.
- Vibration measurement.

Proactive Maintenance:
- ❖ A combination of Preventive and Predictive maintenance.
- ❖ Most expensive and requires skilled and trained personnel.
- ❖ Avoiding unexpected downtime + components last longer →
- ❖ Operational cost is reduced 30% over when only reactive maintenance is considered.

7

7

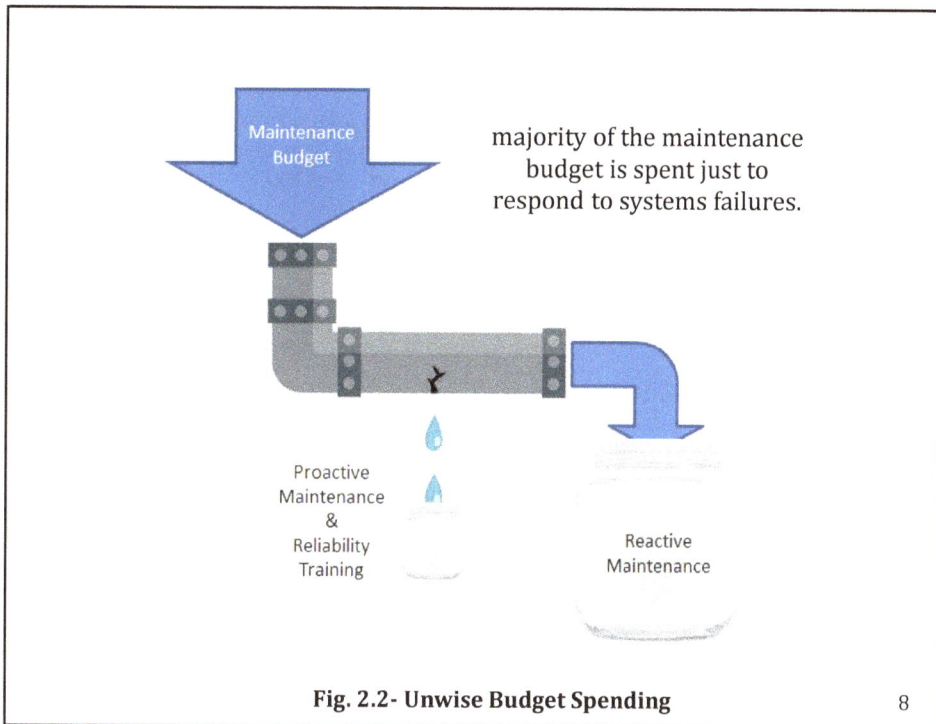

majority of the maintenance budget is spent just to respond to systems failures.

Maintenance Budget

Proactive Maintenance & Reliability Training

Reactive Maintenance

Fig. 2.2- Unwise Budget Spending

8

8

In this textbook:

o   for each hydraulic component, some preventive maintenance actions are suggested and scheduled.

o   The frequency of the suggested actions is based on the best understanding and the experience of the author under assuming that the system is working under normal operating conditions.

o   This frequency may change based on several factors.

o   Therefore, it is highly recommended to review the components and system manufacturer for further instructions.

9

9

2.3- Factors that affect the Frequency of Maintenance

❖ **More frequent checking and inspection in the following cases:**
- High working pressure → effect of contamination is more severe.
- High working temperature → hydraulic fluid accelerated breakdown.
- Highly contaminated work environment → high risk of failure.
   o Example 1: cement factories have a lot of dusty air.
   o Example 2: offshore workplaces have a lot of moisture.
- Very critical application → high risk of life loss
   o Example 1: Aerospace industry.
   o Example 2: Expensive machines and downtime high cost.

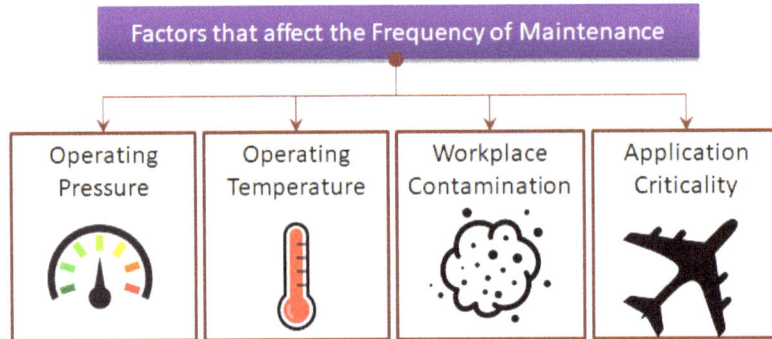

Fig. 2.3- Factors that affect the Frequency of Maintenance

10

10

2.4- Hydraulic Systems Maintenance Record Keeping

Hydraulic Systems Maintenance Record		
Equipment #		
Location		
Service Person	Date	Action Performed

Table. 2.1- Suggested Record Sheet for Hydraulic System Maintenance

11

11

2.5- Skill Set for Maintenance and Troubleshooting Team

The person who is involved in hydraulic system maintenance and/or troubleshooting must maintain a good level of understanding of the following:

- **Safety:** before, during, and after.
- **Basic Principles:** force, work, energy, power, pressure, flow, basic math, etc.
- **Hydraulic Components:** Construction and operating principle.
- **Hydraulic Systems:** Symbols and schematic.
- **Hydraulic Fluids:** Sampling, analysis, and interpretation of analysis report.
- **Filtration:** ISO standards for contamination level.
- **Filters:** Types, application, clogging conditions, cleaning, and replacements.
- **Reservoir:** Cleaning and make up fluid.
- **Plumping:** Types, sizes, cleaning, fittings, replacement, tightening, etc.
- **EH Valves:** Construction, operation, null position, flow gain, and pressure gain.
- **Troubleshooting Principles:** Utilize "Root Cause Failure Analysis".
- **Commissioning:** Safety and startup to manufacturer's instructions.
- **Flushing:** Procedure and evaluation methods.

12

12

2.6- Setting Hydraulic Systems Stages of Services

First Stage of Service (On Spot):
- At the place of the machine.
- Simple maintenance tasks (e.g. filter change, make up oil, line replacement, etc).

Second Stage of Service (Local Repair Shop):
- At a plant or at a dealer's service center.
- Intermediate repair tasks (e.g. pump and cylinder rebuilds).

Third Stage of Service (Manufacturer):
- At the manufacture's site or certified service center.
- Complex repair tasks (e.g. servo valve maintenance and reset, etc.)

Fig. 2.4- Line of Services (Courtesy of Vickers)   13

13

2.7- Common Mistakes during Hydraulic System Maintenance

❖ **Mistake 01-Tightening Torque:**

▪ Tightening to unspecified torque → body distortion + spool seizure + mechanical stress.

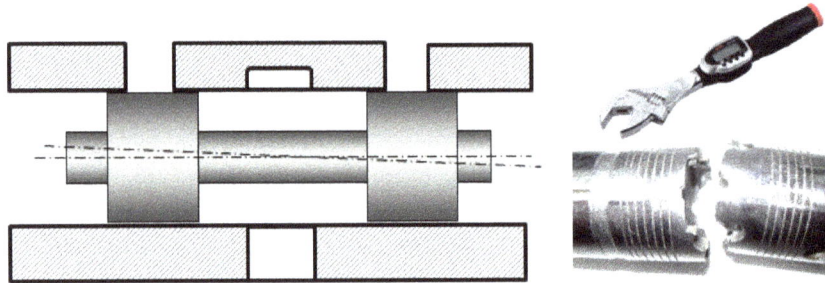

Fig. 2.5- Valve Body Distortion due to Over Tightening Torque

14

14

❖ **Mistake 02-Running a Component on Dry Conditions:**

▪ A pump must never run without oil.
▪ An EH valve if signalized without oil → lack of lubrication → internal clearances affected.

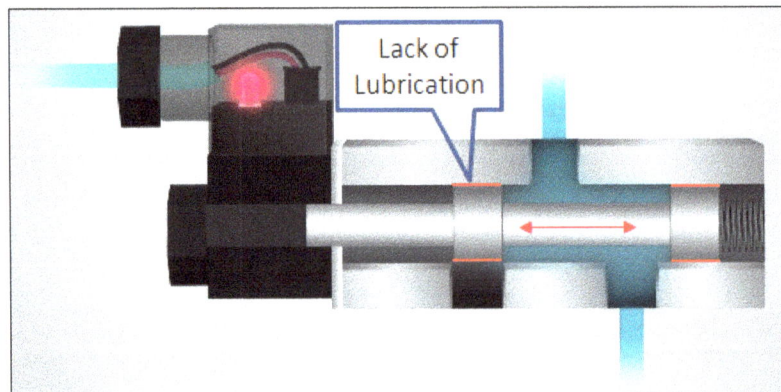

Fig. 2.6- Signalizing a Valve with no Oil affects the Internal Clearances

15

15

❖ **Mistake 03-Use of Inadequate Washing Fluid:**
- Specified Solvents → remove dirt + lubricate moving parts as well.
- Unspecified Solvents (Benzene fuel) →
- incompatible + does not provide lubrication + improper cleanliness level →
- Make it difficult to assemble parts.

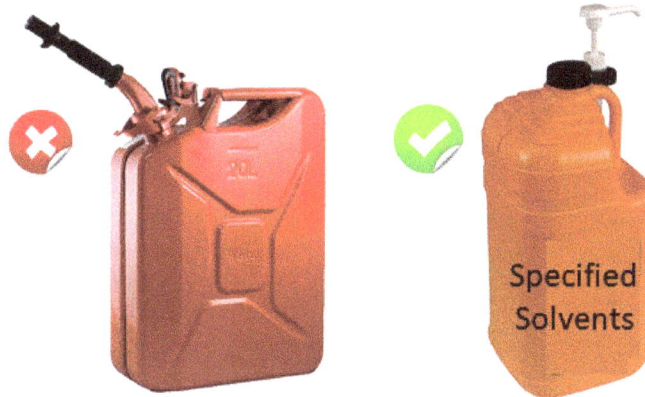

Fig. 2.7- Only Specified Solvents in Cleaning Internal Parts

16

16

❖ **Mistake 04-Use of Inadequate Cleaning Towels:**
- Used for wiping and drying hydraulic components after washing.
- Must be industrially specified and lint-free.

Fig. 2.8- Cleaning Towel

❖ **Mistake 05-Search for Leakage with the Operator's Hand:**
- Never use your hands to check oil leakage → oil injection.
- Alternatively, special tools must be used to search for leakage.

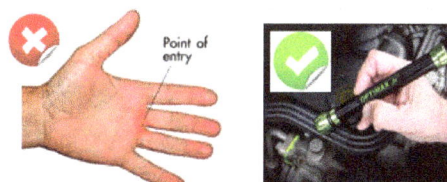

Fig. 2.9- Oil Injection

17

17

❖ **Mistake 06-Improper Line Routing:**
- Improper Routing of transmission lines →
- Increasing energy losses + premature failure.

❖ **Mistake 07-Long hoses Unfastened:**
- *Unfastened* transmission lines → Noise + vibration + rubbing against sharp edges.

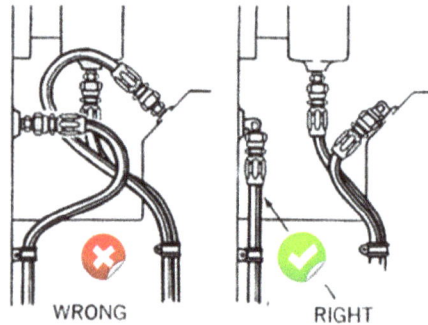

WRONG   RIGHT

Fig. 2.10- Proper Routing of Hoses | **Fig. 2.11- Fastening Long Hoses**

18

18

❖ **Mistake 08-Wrong Identification of a Pump Suction Port:**
- Improperly connected pump → pump cavitation.

❖ **Mistake 09-Wrong Identification of the Direction of Rotation:**
- A pump rotates in wrong direction → pump cavitation.

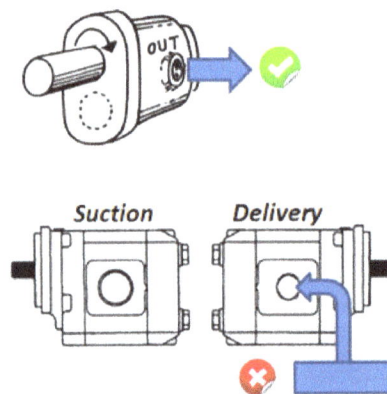

Suction   Delivery

Fig. 2.12- Incorrect Pump Connection   19

19

❖ **Mistake 10-Shutoff Valve on a Pump Suction Line is Left Unlocked:**
▪ A valve on suction line must be locked open during machine operation.

Fig. 2.13- Lockable Shut Off Valve

❖ **Mistake 11-Mixing Different Types of Hydraulic Fluids:**
▪ Mixing different types of hydraulic fluid → unfavorable products.

❖ **Mistake 12-Make Up Oil Without Filters:**
▪ Oil in the reservoir may be cleaner than the unused oil in the barrel.

20

20

❖ **Mistake 13-Changing Component Location:**
▪ Change of component location without circuit analysis →
▪ Possible (change sequence of operation + damage + pressure intensification, etc.)

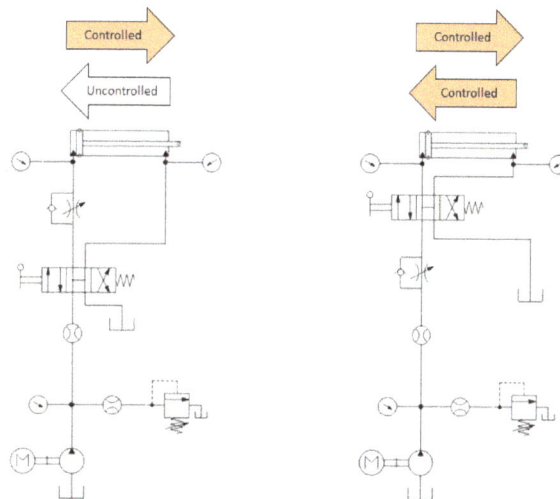

Fig. 2.14- Meter-In Speed Control   21

21

❖ **Mistake 14-Changing Component Specification:**
- Undersizing and oversizing components are prohibited.
- Similar symbols doesn't mean same specifications.
- It is essential to match the part number between old and new component.

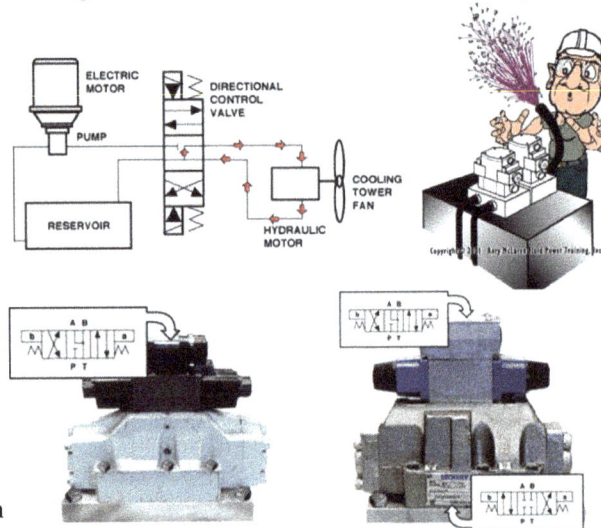

Fig. 2.15- Hydraulic Driven Cooling System

22

22

❖ **Mistake 15-Changing Component Settings:**
- Only trained and authorized personnel are permitted
- Valves with lockable knobs are recommended for critical adjustments.

- **Sequence Valve**: Improper setting → work piece destruction.
- **Pressure Relief Valve:**
  o Setting a PRV Low → inoperative hydraulic actuators.
  o Setting a PRV High → machine or work piece destruction.

Video 157 (1 min)

- **Variable Displacement Pump:**
  o Power-Controlled Pump set Low → loss of power + inoperative system.
  o Power-Controlled Pump set High → prime mover stalls.
  o Pressure-comp. pump set Low → reduced pressure + stopped actuators.
  o Pressure-comp. pump set High → Increased pressure + damage.
  o Displacement–Controlled Pump set Low → reduced flow + slow actuators.
  o Displacement–Controlled Pump set High → increased flow + possible:
  ✓ Turbulent flow in the pipelines.
  ✓ Higher flow forces in the valves.
  ✓ Increased speed of actuators.

23

23

❖ **Mistake 16-Reuse of Sealing Elements:**
▪ Some Sealing Elements, perform their job when they are plastically deformed and squeezed into their installation space.
▪ Reuse such sealing elements → leakage.

Fig. 2.16- Sealing Elements

24

24

❖ **Mistake 17-Incompatible Sealing Material:**
▪ Incompatible sealing element →
▪ Seal physical and chemical deterioration + Internal/external leakage +
▪ Blocking control orifices + stick-slip actuator motion.
▪ The following factors must be considered when choosing a seal material:
   o Temperature Compatibility.
   o Pressure Compatibility.
   o Fluid Compatibility.

❖ **Mistake 18-Improper Pump Priming:**
▪ Improper pump prime → cavitation and premature failure.

❖ **Mistake 19-Improper Air Bleeding:**
▪ Improper air bleeding from a cylinder → sudden or erratic movement.

❖ **Mistake 20-Potential System Contamination and Environmental Pollution:**
▪ Make sure no oil spilling during component disassembling.
▪ Make sure not to leave disassembled component ports uncovered.

25

25

2.8- Reasons to Void Warranty During Hydraulic System Maintenance

- **Reason 1:** Oil contamination level is higher than what is specified by the manufacturer.

- **Reason 2:** Use of unspecified fluid. For example, using regular fluid instead of Extreme-Pressure (EP) fluid that is specified by the manufacturer.

- **Reason 3:** Use of non-original spare parts to recondition a hydraulic component.

- **Reason 4:** Abuse a component by using it out of the recommended maximum operating conditions; namely pressure, temperature, and speed.

- **Reason 5:** Improper installation of a component with disrespect to the manufacturer's instructions.

- **Reason 6:** Setting components that are specified to be set only by the manufacturer or certified agent.

26

26

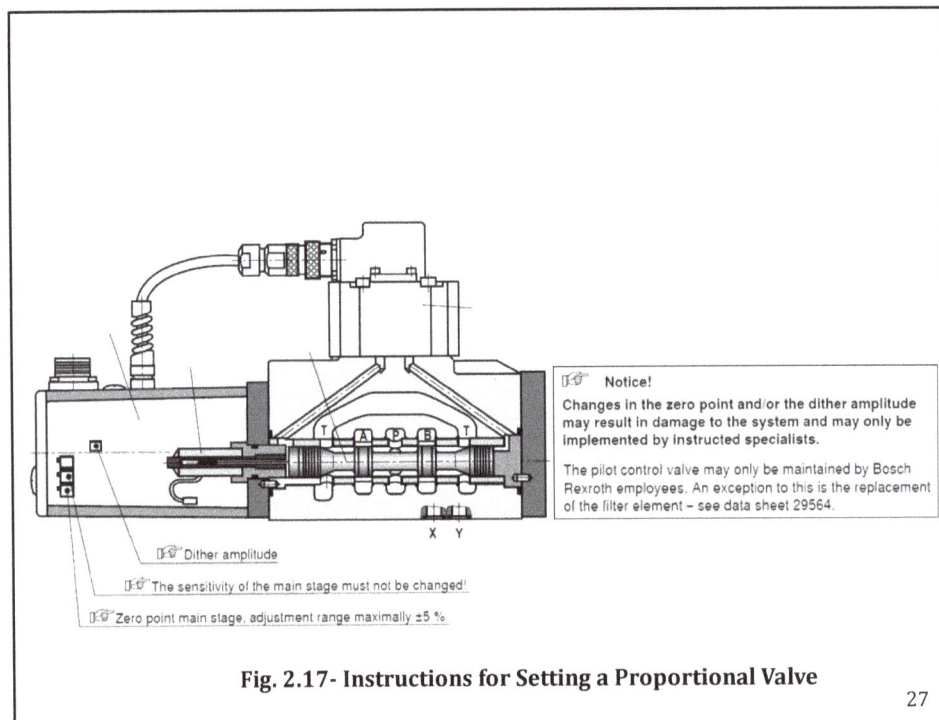

Fig. 2.17- Instructions for Setting a Proportional Valve

27

27

2.9- Standard Tests for Hydraulic Components

2.9.1- Reasons for Tests
- Engineering Validation.
- Product Qualification.
- Quality Assurance at Production.
- Failure Mode Analysis.

2.9.2- Testing Standard Sources
- American Society of Mechanical engineers (ASME).
- Society of Automotive engineers (SAE).
- National Fluid Power Associations (NFPA).
- International Organization for Standardization (ISO/TC 131).
- Application Specific.
  - MIL Standards.
  - Manufacturers.
  - Other Countries.

28

28

Chapter 2 Reviews

1. Which of the following maintenance approaches requires continuous monitoring of the system conditions and in-depth analysis of the system performance?
   A. Reactive maintenance.
   B. Preventive maintenance.
   C. Predictive maintenance.
   D. None of the above.

2. Lack of maintenance of a hydraulic system leads to?
   A. Unsafely working system.
   B. Leaking and inefficient system.
   C. Noisy and less productive system.
   D. All of the above.

3. Factors that affect the frequency of preventive maintenance actions are?
   A. Operating pressure, operating temperature, workplace contamination, and application criticality.
   B. Viscosity and the acidity in hydraulic fluids.
   C. Size and power of the pump.
   D. All of the above.

4. Which of the following liquids are allowed for cleaning hydraulic components?
   A. Same hydraulic fluid used in the component.
   B. Benzene.
   C. A solvent that is specified by component/system manufacture and cleaned to standard.
   D. Hot Water.

5. One of the common mistakes during hydraulic system maintenance?
   A. Only trained person performs the maintenance.
   B. Always start by checking the oil level.
   C. Reuse of O-Rings and Cutting Rings.
   D. Replacing filter element before fully clogged.

Chapter 2 Assignment

Student Name: --- Student ID: ------------------

Date: -- Score: -----------------------

Question: List the set of skills that a maintenance and troubleshooting person should have a good level of understanding about them?

Chapter 3
Hydraulic Measuring Instruments

Objectives:

This chapter provides an overview of the common measuring devices used in hydraulic systems including devices for measure pressure, flow, temperature, oil level, and load cells. The chapter introduces the difference between a meter, a switch, and a sensor. The chapter also discusses the best practices for measuring devices selection & replacement, maintenance scheduling, installation & maintenance, and standard tests & calibration.

0

0

Brief Contents:

1

1

3.1- Classification of Measuring Instruments

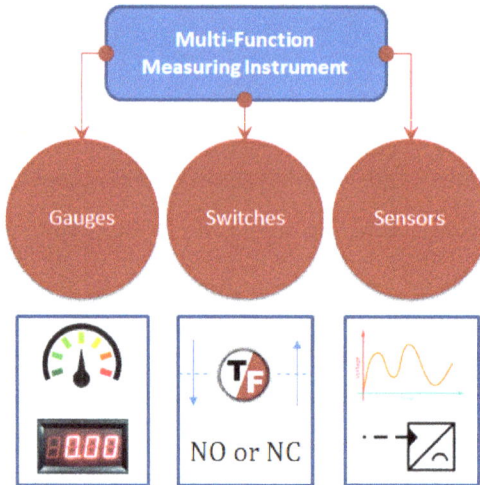

Fig. 3.1- Classification of Measuring Instruments

2

2

3.2- Pressure Measuring Instruments
3.2.1- Pressure Gauges

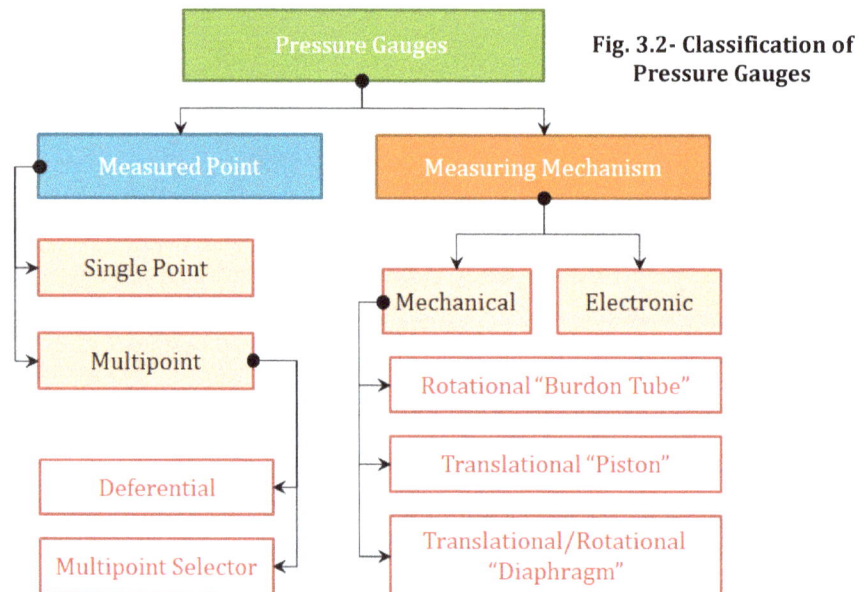

Fig. 3.2- Classification of Pressure Gauges

3

3

3.2.1.1- Burdon Tube Pressure Gauge

V36 (0 min)

- Single point pressure measuring.
- Rotational mechanism.
- Working mechanism is specially constructed and calibrated.

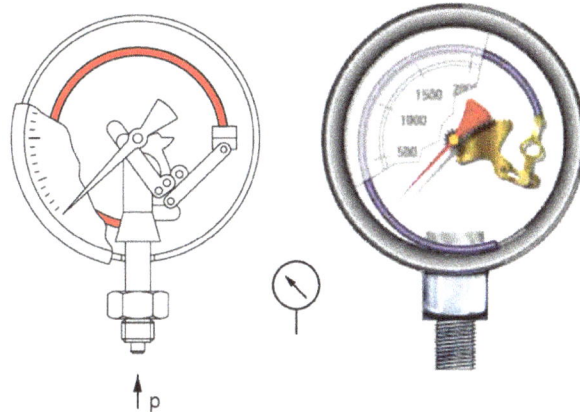

Fig. 3.3- Burdon Tube Pressure Gauge (Courtesy of Rexroth)

4

4

3.2.1.2- Piston-Type Pressure Gauge

- Single point pressure measuring.
- Translational mechanism.
- Spring stiffness is calibrated to read pressure.

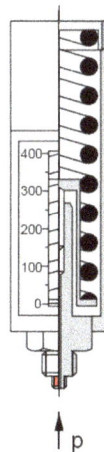

Fig.3.4- Piston-Type Pressure Gauge (Courtesy of Rexroth)

5

5

3.2.1.3- Diaphragm-Type Pressure Gauge

- Single point pressure measuring.
- Translational –Rotational mechanism.
- Diaphragm deflection is calibrated to read pressure.

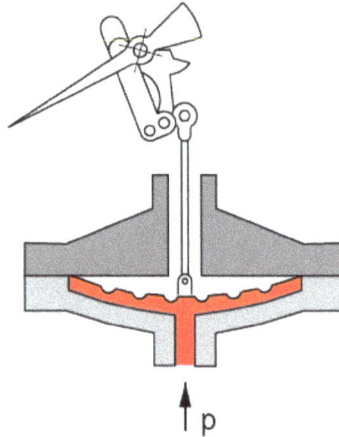

Fig. 3.5- Diaphragm Type Pressure Gauge (Courtesy of Rexroth)

6

6

3.2.1.4- Pressure Gauge with Digital Readings

- Single point pressure measuring.
- Working mechanism (strain gauge) $\rightarrow$ electrical signal.
- Diaphragm deflection is calibrated to read pressure.

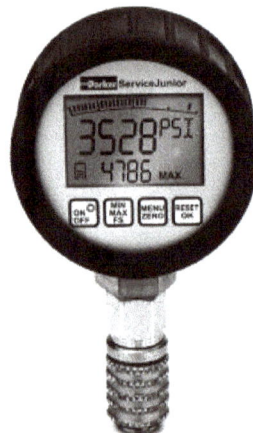

Fig. 3.6- Pressure Gauge with Digital Reading (Courtesy of Parker)

7

7

3.2.1.5- Differential Pressure Gauge

- Measures the difference in pressure between two points.
- Used for diagnosing proper functioning of hydraulic components.
- Example (clogging condition of a filter).
- Pressure reading (dial or digital)

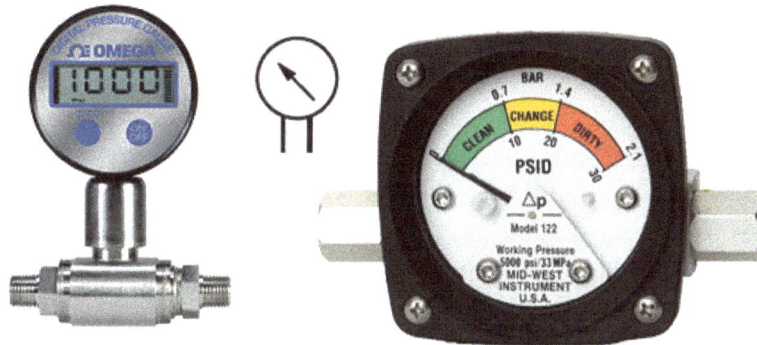

Fig. 3.7- Differential Pressure Gauge

8

3.2.1.6- Multipoint Pressure Gauges

- Measures pressure at more than two points.
- Rotatable knob is used to select one reading at a time.
- Saves space for installing multiple gauges.

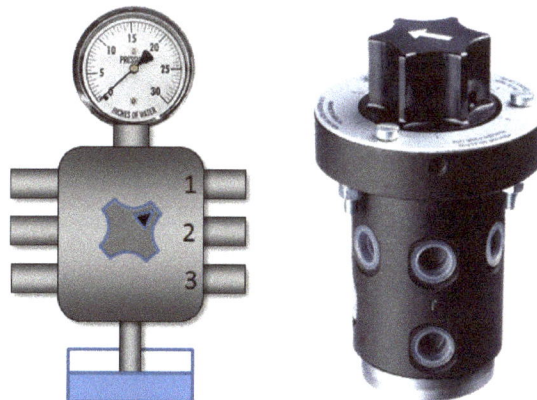

Fig. 3.8- Multipoint Pressure Gauge

9

3.2.2- Pressure Switches

- On-Off control mode - binary electrical signal - (TRUE or FALSE).
- When raising or falling.

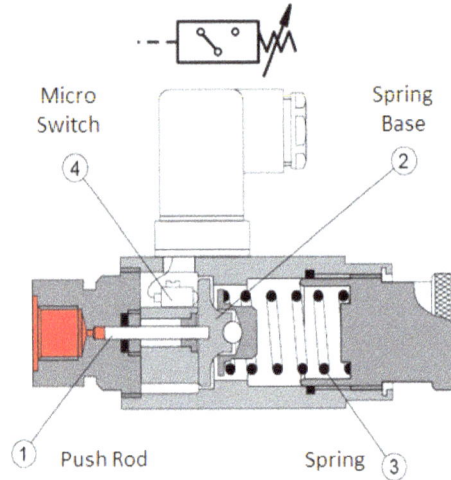

Fig. 3.9- Normally-Closes Pressure Switch (Courtesy of Rexroth)

10

10

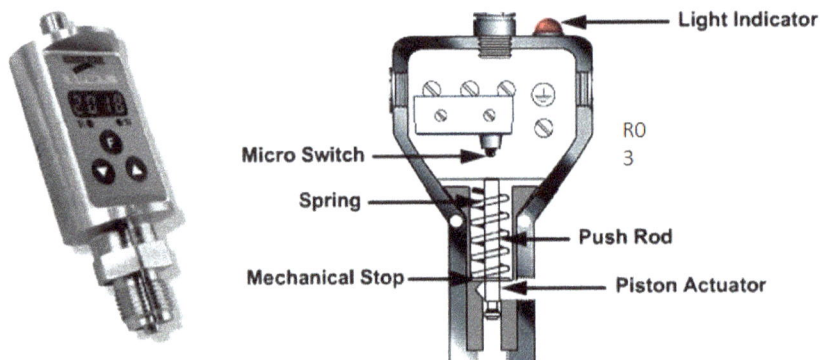

Fig. 3.10- Normally-Open Pressure Switch

11

11

3.2.3- Pressure Sensors

- Continuous control mode.

- Transducer converts pressure into a proportional electrical.

- Signal format (voltage + current).

Fig. 3.11- Examples of Pressure Sensors

12

Specification	
TR-PS2W-100BAR	
Measuring range	**Resolution**
0 ... 100 bar	0.1 bar
0 ... 1450 PSI	2 psi
0 ... 101.9 Kg / cm²	0.1 kg / cm²
0 ... 75000 mm Hg	100 mm Hg
0 ... 2952 inch Hg	2 inch Hg
0 ... 1019 meters H20	1 meter H20
0 ... 40100 inches H20	50 inches H20
0 ... 98.7 ATP	0.1 ATP
0 ... 10000 kPa	10 kPa
Exit	4 to 20-mA DC
4-mA	= normal pressure
20-mA	= maximum pressure
Precision pressure sensor made of ceramic	
Linearity output signal	± 1% FS
Zero Offset	± 2% FS
Operating temperatur	-20 ... 80°C / -4 ... 176°F
Ambient humidity	80% RH
Size	Ø 30 mm x 85 mm
Weight	160 g / < 1 lb

13

3.2.4-Multifunction Pressure Measuring Instruments

- All-in-one device (pressure gauge, pressure switch, and pressure sensor).

- A fully programmable device offers (local display + electrical outputs).

- It provide critical feedback to a control system.

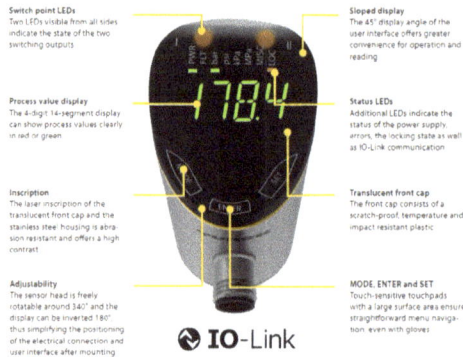

Fig. 3.12- Multifunction Programmable Pressure Sensor (Courtesy of Turck)   14

14

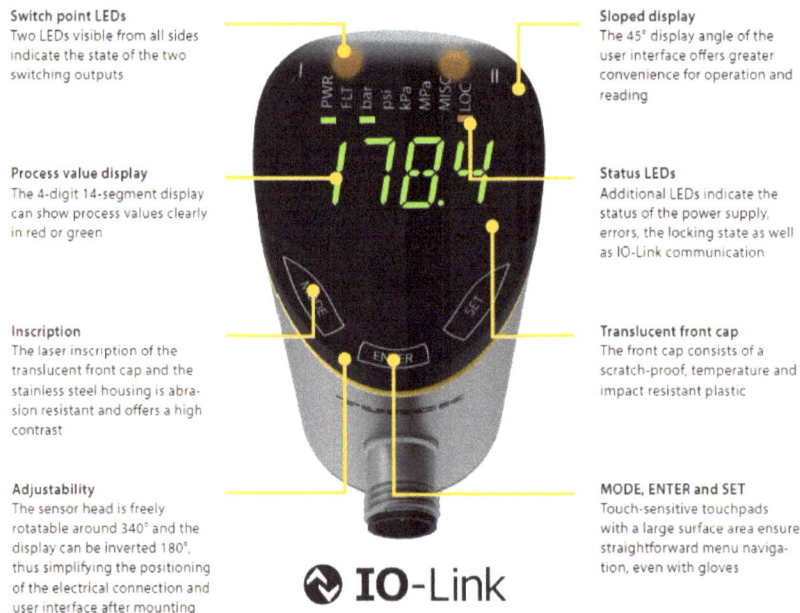

15

3.3- Flow Measuring Instruments
3.3.1- Operating Principles for Flow Measurement

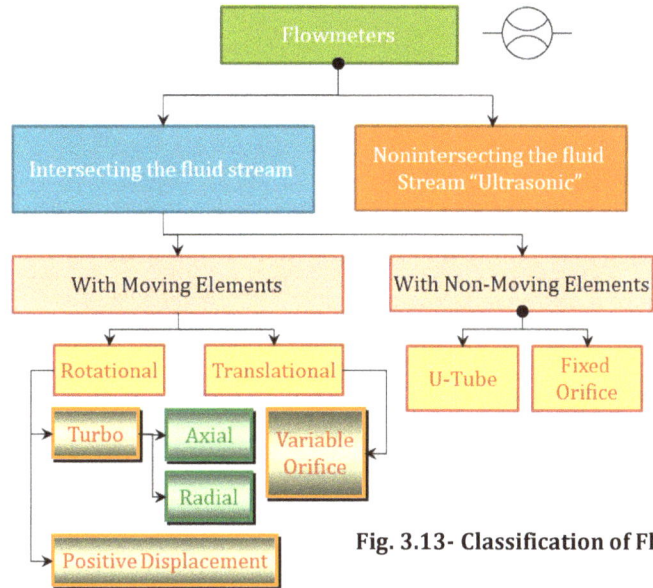

Fig. 3.13- Classification of Flowmeters

16

16

3.3.2- Fixed Orifice Flowmeters, Switches, and Sensors

- Correction factors based on the density and viscosity of the fluid.

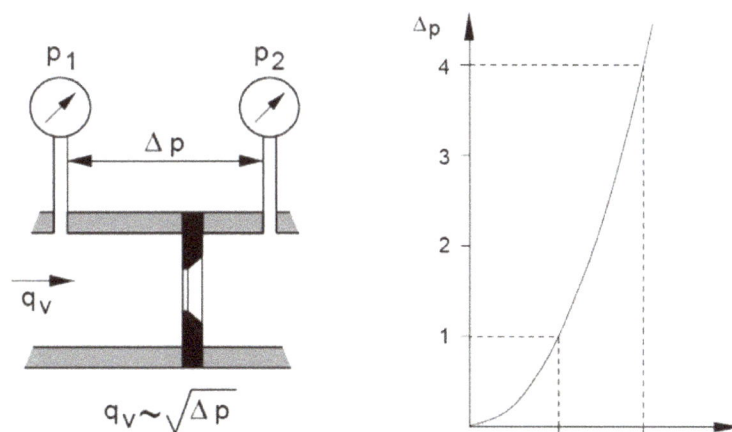

$$q_v \sim \sqrt{\Delta p}$$

Fig. 3.14.A- Operating Principle of Fixed Orifice Flowmeters

17

17

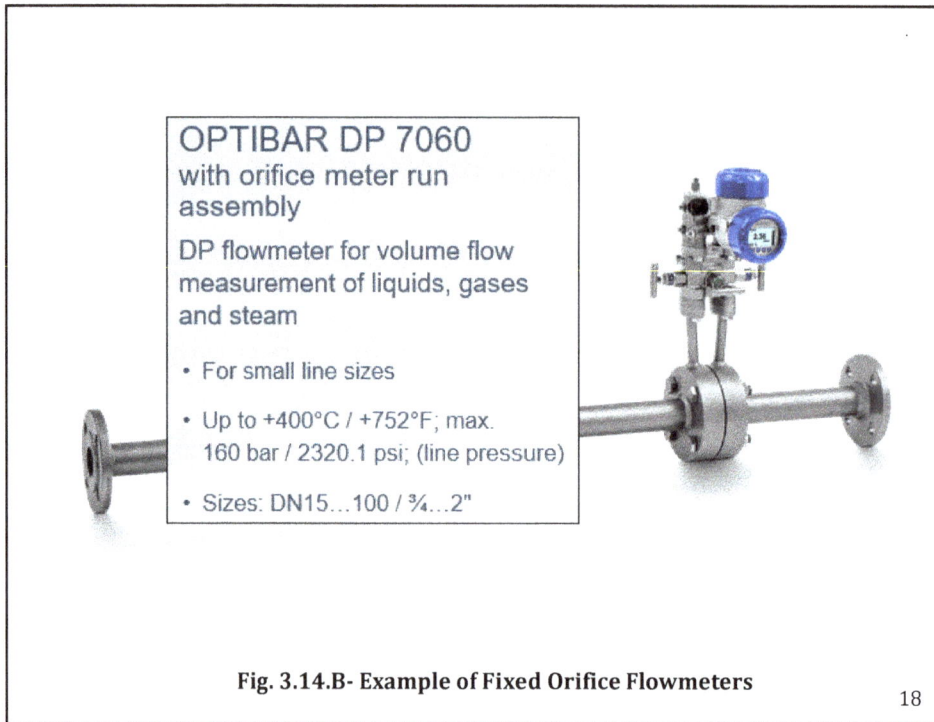

OPTIBAR DP 7060

with orifice meter run assembly

DP flowmeter for volume flow measurement of liquids, gases and steam

- For small line sizes

- Up to +400°C / +752°F; max. 160 bar / 2320.1 psi; (line pressure)

- Sizes: DN15...100 / ¾...2"

Fig. 3.14.B- Example of Fixed Orifice Flowmeters

18

18

3.3.3- U-Tube Flowmeters, Switches, and Sensors

- Correction factors based on the density and viscosity of the fluid.

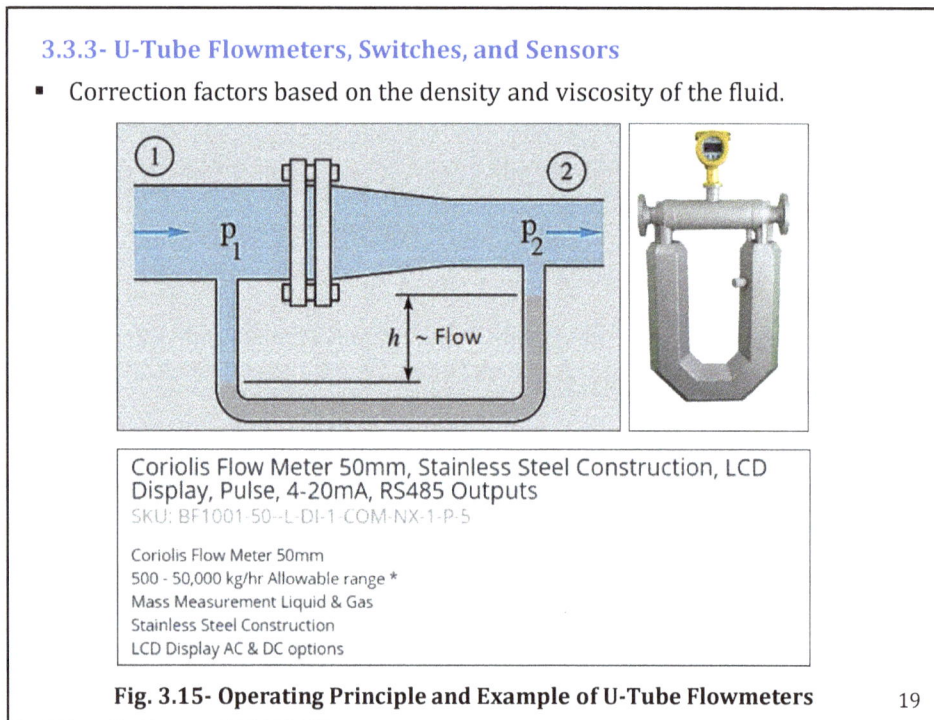

Coriolis Flow Meter 50mm, Stainless Steel Construction, LCD Display, Pulse, 4-20mA, RS485 Outputs
SKU: BF1001-50--L-DI-1-COM-NX-1-P-5

Coriolis Flow Meter 50mm
500 - 50,000 kg/hr Allowable range *
Mass Measurement Liquid & Gas
Stainless Steel Construction
LCD Display AC & DC options

Fig. 3.15- Operating Principle and Example of U-Tube Flowmeters

19

19

3.3.4- Variable Orifice Flowmeters, Switches, and Sensors

- Variable Orifice Flowmeter is also known as Float Flowmeter.
- It is a one-way flowmeter.
- Fluid flow enters the meter → pressure created around the cone →
- Piston moves linearly proportional to flow .
- Accuracy is (low-medium) ± (2-5) %.

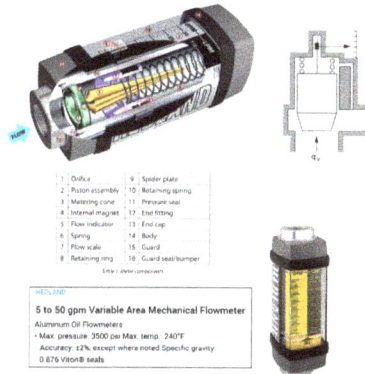

Fig. 3.16- Operating Principle and Example of Variable Orifice Flowmeters (Courtesy of Hedland)

20

20

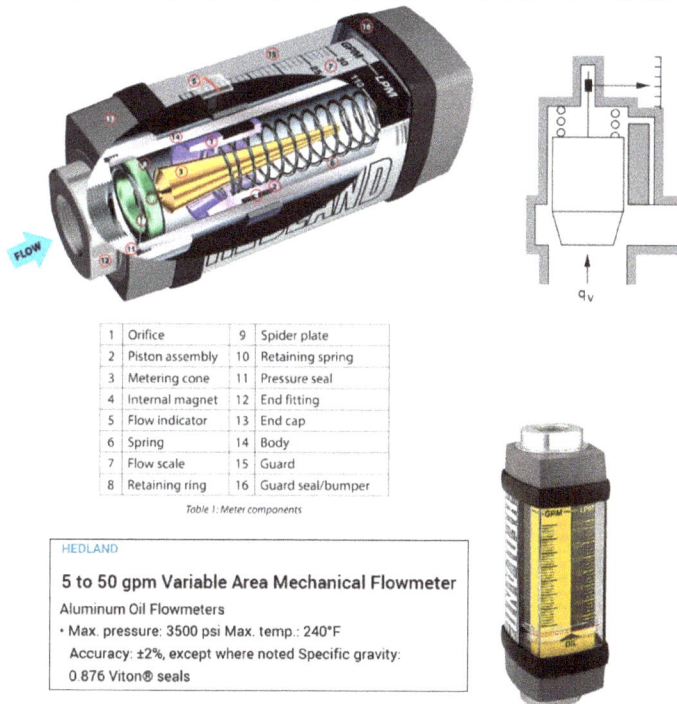

1	Orifice	9	Spider plate
2	Piston assembly	10	Retaining spring
3	Metering cone	11	Pressure seal
4	Internal magnet	12	End fitting
5	Flow indicator	13	End cap
6	Spring	14	Body
7	Flow scale	15	Guard
8	Retaining ring	16	Guard seal/bumper

Table 1: Meter components

HEDLAND

5 to 50 gpm Variable Area Mechanical Flowmeter

Aluminum Oil Flowmeters
- Max. pressure: 3500 psi Max. temp.: 240°F
  Accuracy: ±2%, except where noted Specific gravity:
  0.876 Viton® seals

21

21

- Another type of variable orifice flowmeter.
- It is a one-way flowmeter + built in thermometer for temperature reading.
- A sharp-edged orifice and tapered metering piston.
- Piston movement is proportional to the flow rate.
- Sharp edge orifice minimizes the effects of viscosity → better accuracy.

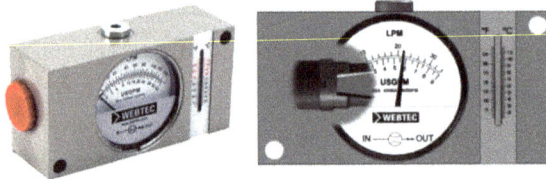

Specification

- Ambient temperature: -10 to 50°C (14 to 122°F)
- Fluid temperature:
  Continuous: 20 to 80°C (65 - 176°F)
  Intermittent use (less than 10 minutes) > 80 to 110°C (> 176 - 230°F)
- Humidity: 10 - 90 % RH
- Fluid: see model configuration
- Seals: FKM as standard (EPDM available on request)
- Accuracy: 4% of full scale (calibrated at 28 cSt)
- Pressure: 420 bar (6000 psi)
- Flow range: see model configuration

Fig. 3.17- Operating Principle and Example of Variable Orifice Flowmeters (Courtesy of Webtec)

22

22

3.3.5- Turbine Flowmeters, Switches, and Sensors

- Axial or Radial Turbine flowmeters.
- Flow is proportional to speed of the turbine.
- Accuracy (medium to high) with error of <2%.

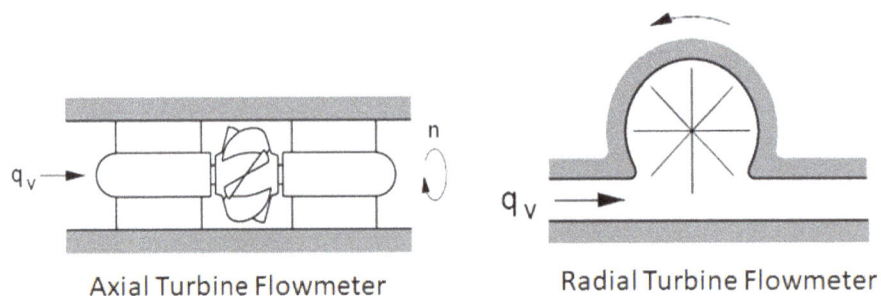

Axial Turbine Flowmeter

Radial Turbine Flowmeter

Fig. 3.18- Operating Principle of Turbine Flowmeters

23

23

INTERNATIONAL HYDAC

Flow Rate Transmitter
HFT 3100 Ex applications

| Turbine | High accuracy | Additional measuring connections |

Technical data:

Input data	
Measuring range and operating pressure	1.2 .. 20.0 l/min 420 bar
	6.0 .. 60.0 l/min 420 bar
	15.0 .. 300.0 l/min 420 bar
	40.0 .. 600.0 l/min 420 bar
Additional connection options [1)]	2x G 1/4 female threads for pressure or temperature sensors with relevant approvals
Housing material	Stainless steel 1.4404
Parts in contact with fluid	Stainless steel: 1.4404, 1.4460, tungsten carbide
Output data	
Output signal, permitted load resistance	4 .. 20 mA, 2-conductor, with HART protocol $R_{Lmax} = (U_B - 12\ V) / 20\ mA\ [k\Omega]$ for HART communication min. 250 Ω
	HART communication acc. to HART 7 specifications
	HART Common Practice Commands, e.g. altering of measuring range limits (see table)
Accuracy	≤ 2 % of the actual value

Fig. 3.19- Example of Turbine Flow Sensors (Courtesy of Hydac)

24

24

- Special and compact design → installed where space is limited.
- Various outputs (4 - 20 mA current loop, 0 - 5 V or 0 - 3 V).
- The turbine blade design → minimized effects of temp. & viscosity.
- Flow straightener before the turbine → reduce flow turbulence → better accuracy

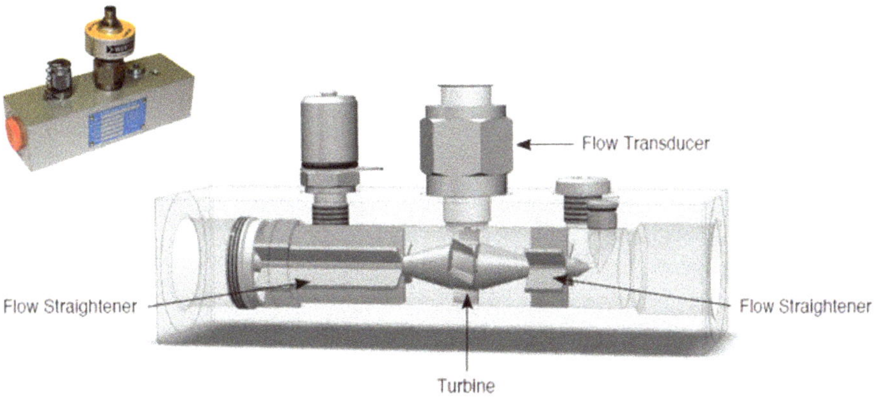

Fig. 3.20- Example of Axial Turbine Flow Sensors (Courtesy of Webtec)

25

25

Cartridge Flow Transmitter:

- ❑ Standard cartridge style design with turbine flow accuracy.
- ❑ Easy integration and no additional hardware or connections are required.
- ❑ Real-time data for predictive maintenance and remote troubleshooting.
- ❑ Continuous flow monitoring of all critical hydraulic functions.
- ❑ Configurable for standard 4-20mA output.

V638 (1 min)

Fig. 3.21- Example of Axial Turbine Flow Sensors
(www.dgdfluidpower.com)

26

26

3.3.6- Positive Displacement Flowmeters, Switches, and Sensors

- ▪ Flow is proportional to the rotational speed.
- ▪ Most accurate flow meters.
- ▪ Used for purpose of flow/speed automatic control.

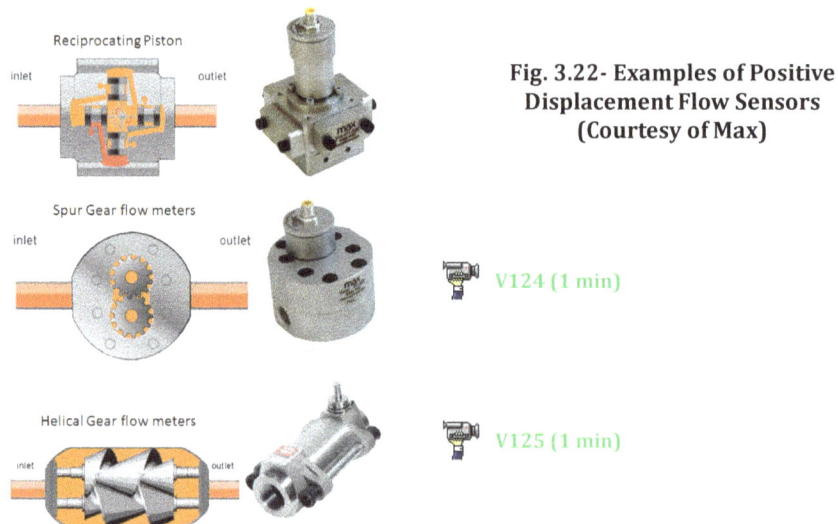

Fig. 3.22- Examples of Positive Displacement Flow Sensors (Courtesy of Max)

V124 (1 min)

V125 (1 min)

27

27

3.4- Oil Level Measuring Instruments

1. Visually indicator (plus temperature measurement).
2. Oil Level Switches for monitoring and on-off control mode.
3. Oil Level Sensors for remote condition monitoring and purposes of automatic control.

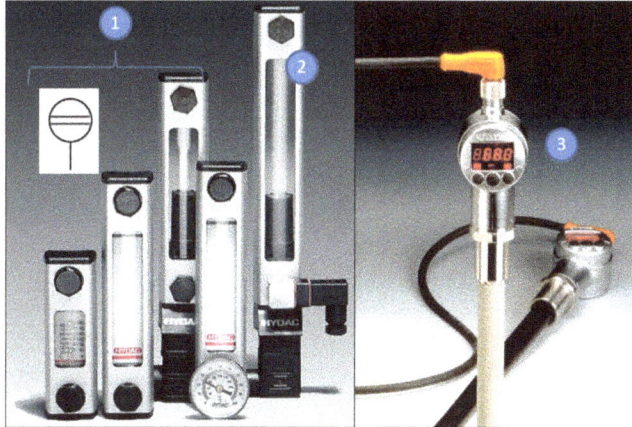

Fig. 3.23- Examples of Oil Level Indicators, Switches, and Sensors (Courtesy of Hydac)

28

28

Other example of fluid level measuring devices as follows:

1. Visual and electrical level indicators.
2. Electrical float level indicator.

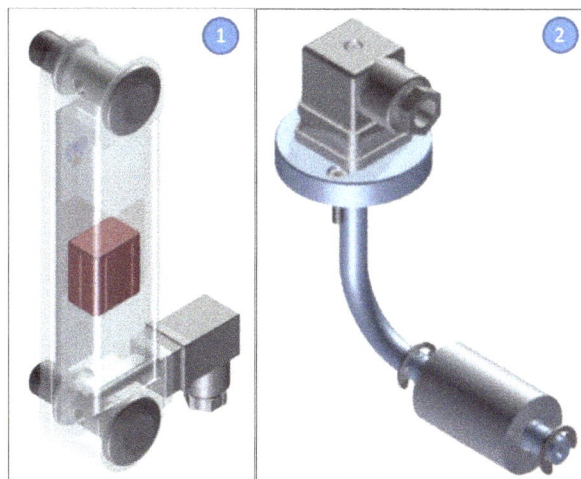

Fig. 3.24- Examples of Oil Level Indicators, Switches, and Sensors (Courtesy of MPFiltri)

29

29

3.5- Oil Temperature Measuring Instruments

1. Visually indicator.
2. Temperature switch for on-off control mode.
3. Temperature switch for remote condition monitoring and purposes of automatic control.

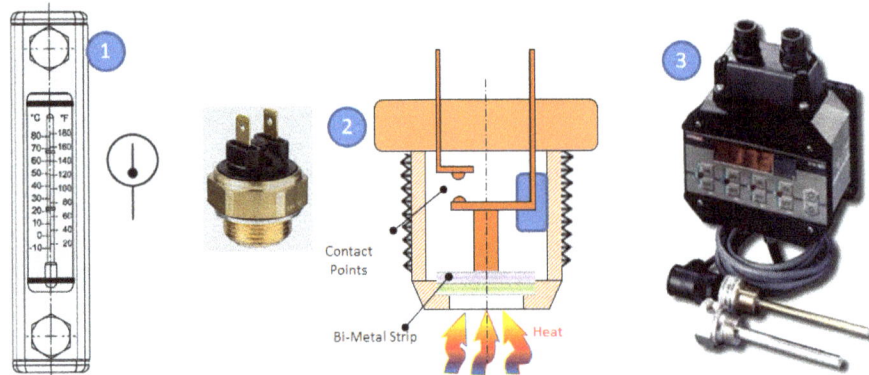

Fig. 3.25- Examples of Oil Temperature Indicators, Switches, and Sensors

30

30

3.6- All-In-One Measurement Instruments

- Portable measuring instruments are available.
- For in field measurement and troubleshooting purposes.

Benefits & Features

- Flow ranges from 0.4 - 7.0 GPM (1.5 - 26 LPM) up to 8 - 160 GPM (30 - 605 LPM)
- Port sizes from SAE 8 (G 1/4) to SAE 20 (G 1-1/4)
- Available with flow, pressure, and temperature sensors in one block
- Flow transducer available with 4-20 mA or 0-5 Vdc output signal
- Temperature and pressure transducers transmit 4-20 mA output signals
- Rated to 5800 PSI (400 Bar)
- 5-point standard calibration, 10 point available
- Flow straighteners manufactured into meter
- Accuracy ±1% of reading @ 32 cSt
- Use with Flo-tech F6700 / F6750 Series Displays

Fig. 3.26- All-In-One Flow, Pressure and Temperature Sensor (Courtesy of Flo-Tech)

31

31

3.7- Load and Torque Cell

- Used for measuring axial loads, bearing forces, and shaft torques.

SPECIFICATIONS
NOMINAL DIAMETER: 20 cm2
LOAD CELL HOUSING MATERIAL: Stainless steel
PISTON: Stainless steel
CONNECTING LINE: Direct connection – standard;
flexible tubing, capillary restrictor
RANGES: From 300 lb_f through 22,000 lb_f

MEASURING INSTRUMENT
PRESSURE GAUGE: 2-1/2" 300 Series,
one piece die cast brass case; dry or liquid filled;
4" 901 Series stainless steel case; dry or liquid filled
TRANSDUCER: 100, 200 or 615 Series transducer
OUTPUT SIGNALS: 4 mA to 20 mA, 2-wire:
0 Vdc to 5 Vdc, 0 Vdc to 10 Vdc, 1 Vdc to 5 Vdc,
1 Vdc to 6 Vdc & 1 Vdc to 11 Vdc, 3-wire
ACCURACY: ±0.5% full scale (BFSL) to ±0.125%
full scale
OPERATING TEMP.: -4 °F to 140 °F (-20 °C to 60 °C)
AMBIENT TEMP.: -4 °F to 140 °F (-20 °C to 60 °C)

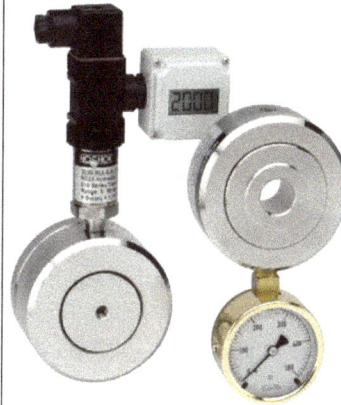

Fig. 3.27- Hydraulic Load Cells (https://www.noshok.com)

32

32

3.8- Other Measuring Devices

- Speed.
- Vibration
- Contamination, etc. V418 (3.5 min)
- These sensors are out of scope of this textbook.

33

33

3.9- Hydraulic Data Logger

❖ Portable multi-function measuring instrument.

❖ Used for testing and evaluation tasks.

▪ **General Functions:** Measuring, monitoring, analyzing and saving data.

▪ **Measured Parameters:** Measuring pressure, temperature, and flow.

▪ **Number of Channels:** Up to 54 channels or up to 26 sensors.

▪ **Sensors Connections:** CAN-bus networks + Standard analogue inputs.

▪ **Sensors Parameterization:** Units and measuring ranges.

▪ **Digital I/O:** One digital input and one digital output are available.

▪ **Display:** various (numeric, bar graph, indicator gauge or curve chart).

▪ **Memory:** built in + microSD cards + USB drives.

▪ **Storage:** Up to 4 million individual values. + one billion total.

▪ **Network:** Ethernet network + LAN ports.

▪ **Software:** PC-based analysis software for data analysis and visualization.

▪ **Safety:** All ports on the instrument are covered with rubber caps to protect them from being touched and from dust and moisture.

34

34

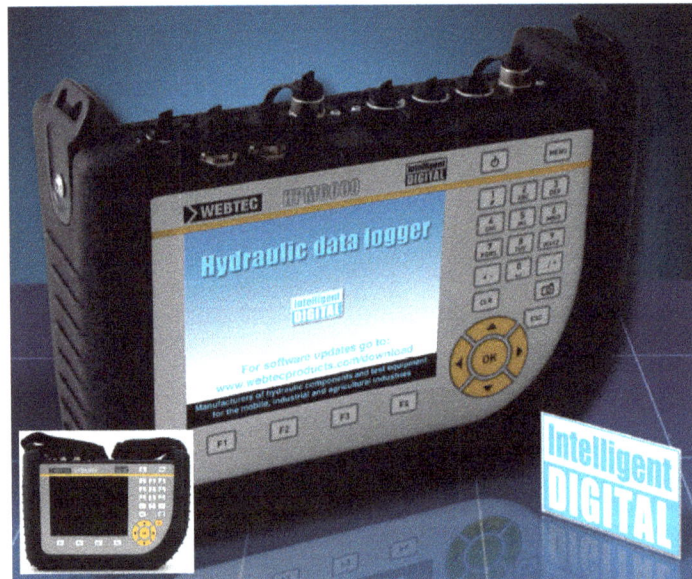

Fig. 3.28- Hydraulic Data Logger (Courtesy of Webtec)

35

35

3.10- Wireless Sensors and IoT in Hydraulic System Maintenance
3.10.1- Traditional versus Wireless Sensors

Traditional Sensors:

- Use reliable wired sensors.

- Creates potential risk for workers in case of visual readings.

- Costs for setting up and wiring.

Wireless Sensors and IoT-Enabled Solutions:

- No risk of visual readings.

- Reduce the cost of setting and wiring.

- Remote condition monitoring from anywhere in the world.

- Smart predictive maintenance and machine diagnosis.

- Reduce unplanned downtime.

- Battery powered wireless sensors require frequent battery charge.

- Cyber security is required for the safety of a machine operation.

36

36

Wireless sensors are available with the following common features:
- User-definable measurement units (bar/psi), (F°/C°), etc.
- User-selectable broadcast intervals.
- Port Options: Male NPTF, SAE, and others.
- Corrosion-resistant materials for challenging environments.

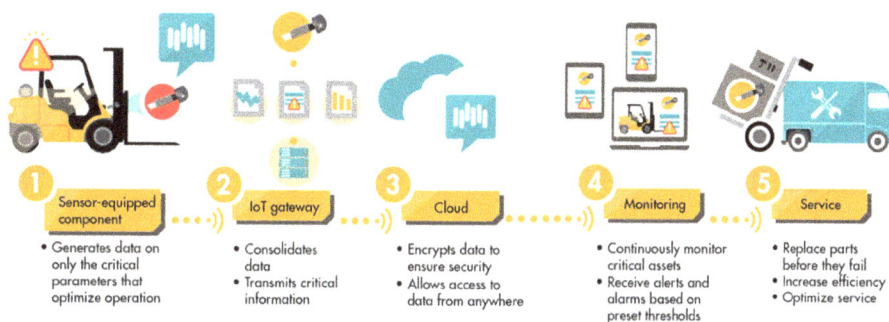

Fig. 3.29- IoT-Enabled Solutions in Industry (Courtesy of Parker)

37

37

3.10.2- Components of Wireless Sensors and IoT-Enabled Solutions

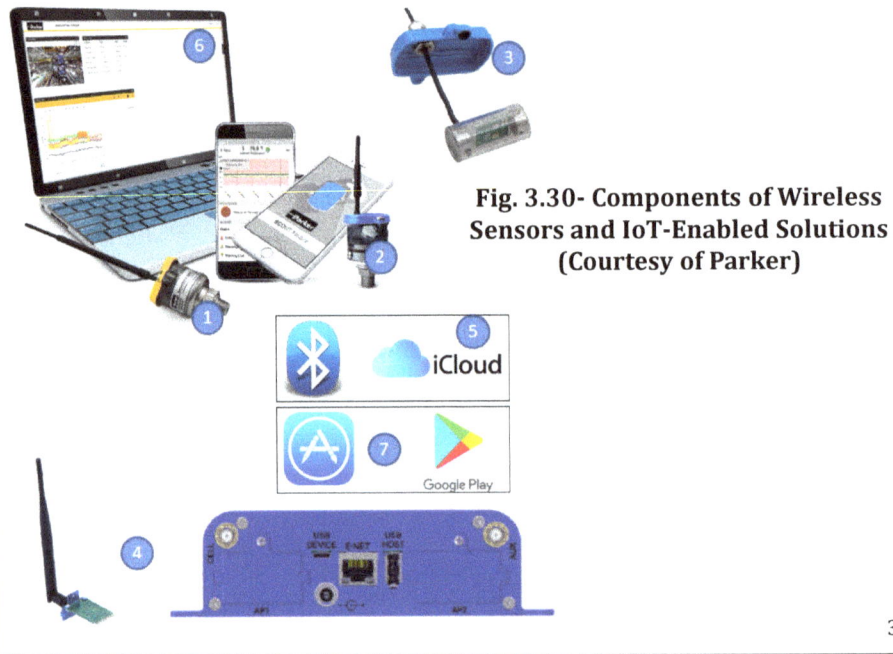

Fig. 3.30- Components of Wireless Sensors and IoT-Enabled Solutions (Courtesy of Parker)

38

38

3.10.3- Wireless Sensors

pressure (1), temperature (2), humidity (3), flexible distance (4), vacuum (5).

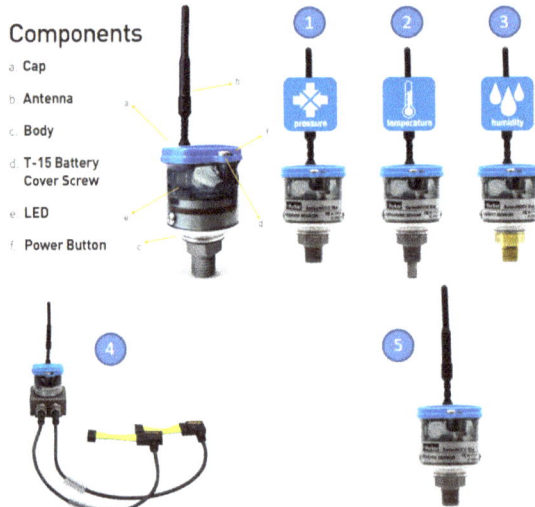

Fig. 3.31- Examples of Wireless Sensors (Courtesy of Parker)

39

39

3.10.4- Wireless Transmitters

- Wired sensor are upgraded by connecting them to a wireless transmitter.
- Customers to receive the information on their mobile device.

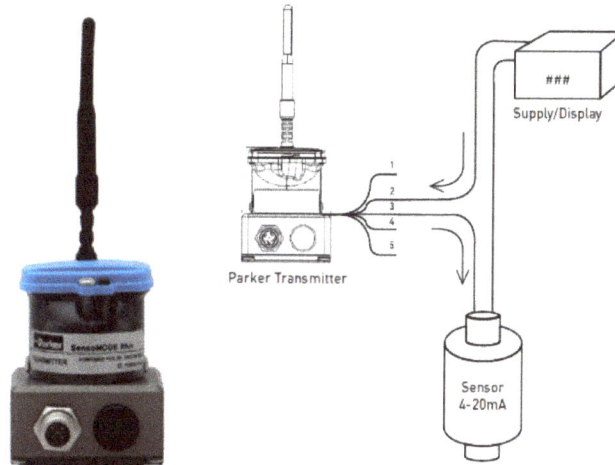

Supply/Display

Parker Transmitter

Sensor
4-20mA

Fig. 3.32- Example of Wireless Transmitter (Courtesy of Parker) 40

40

3.10.5- Power Suppliers for Wireless Sensors

Instead of battery-powered operation of wireless sensors, portable power suppliers are available with the following features:

- Supplies continuous power to sensors and eliminates the need for battery replacement.
- Used with IEC/UL 508 Class 2 power supply.
- Easy upgrade.
- Increases ambient working temperature range of the sensor

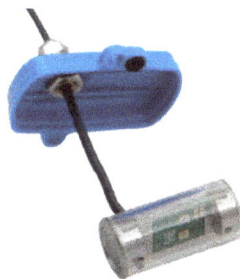

**Fig. 3.33- Example of Power Supply for Wireless Sensors
(Courtesy of Parker)** 41

41

3.10.6- Mobile-Based Connectivity with Wireless Sensors

Mobile Application allows users to:

- Work under iOS and Android.
- Connect to multiple sensors concurrently.
- Offer immediate and historic trend information with analytical tools.
- Sensor setup and alarm setting.

**Fig. 3.34- Example of Mobile Connectivity for Wireless Sensors
(Courtesy of Parker)**

42

42

Example of applying Mobil-based connectivity for Pick & Place location.

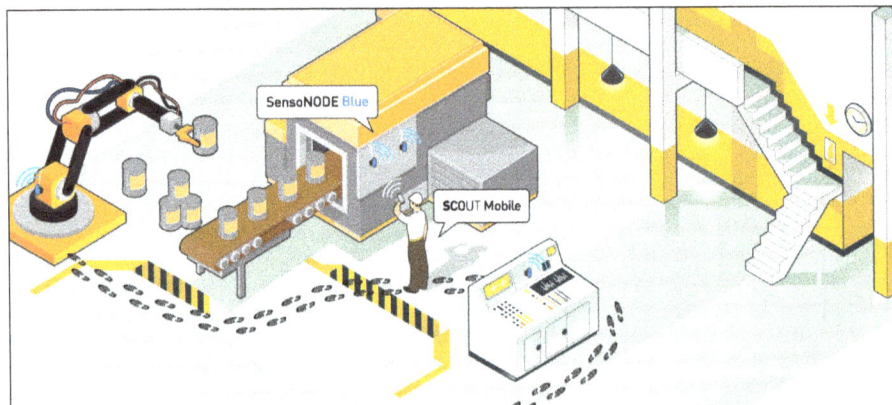

**Fig. 3.35- Example of Mobile-Based Connectivity for Wireless Sensors
(Courtesy of Parker)**

43

43

3.10.7- iCloud-Based Connectivity with Wireless Sensors

❖ iCloud-Based connectivity is more adequate for large operation.

❖ <u>Common features:</u>

▪ Multiple user access data anytime, anywhere.

▪ Export and share data.

▪ Easy web-based interface, no software to download, and no updates.

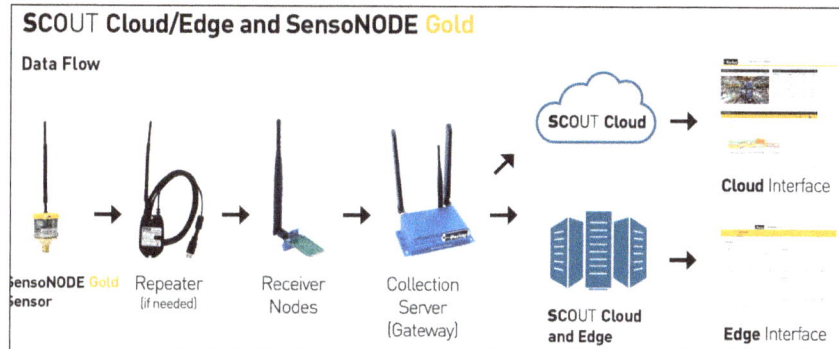

SCOUT Cloud/Edge and SensoNODE Gold

**Fig. 3.36- Components for iCloud Connectivity for Wireless Sensors
(Courtesy of Parker)**

44

44

Examples of applying iCloud-based connectivity for various places simultaneously from one location:

V676 (1.5 min)

Material handling

Injection molding machine

Mining location

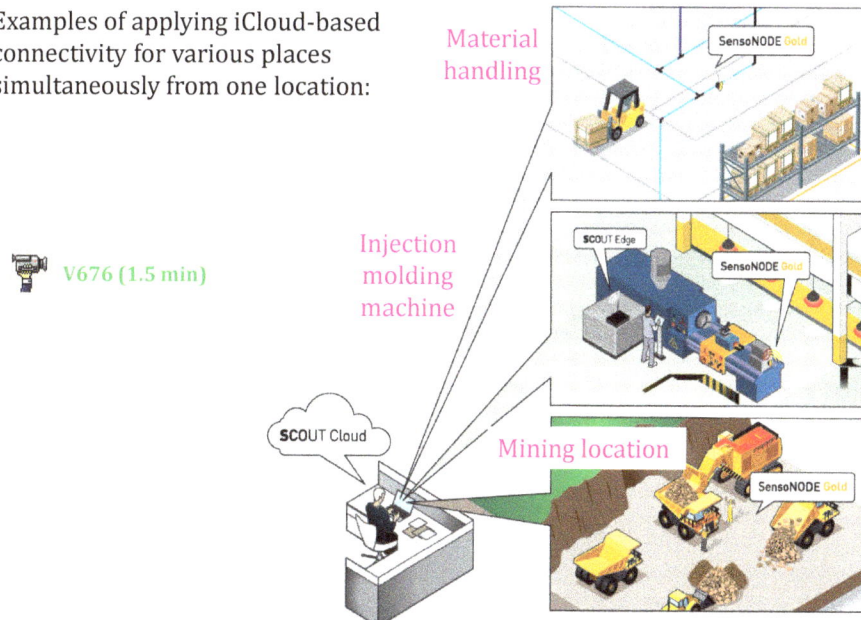

**Fig. 3.37- Examples of iCloud-Based Connectivity for Wireless Sensors
(Courtesy of Parker)**

45

45

3.11- BP-Measuring Instruments-01-Selection and Replacement

When selecting or replacing an existed measuring instrument →

<u>Review:</u>

- Type/size of hydraulic connections to avoid using adaptors.
- Measuring scale and make sure the new one has the same scale.
- Electrical connections and make sure to use standard connections.
- **Switches:** status (NO or NC) and the adjustment range.
- **Sensors:** Output signal (Voltage and Current, 0-20 mA or 4-20 mA).

46

46

3.12- BP-Measuring Instruments-02-Maintenance Scheduling

Unless otherwise stated by components and systems manufacturer:

#	Preventive Maintenance Actions	Daily	Weekly	Monthly	Biannually	Annually
1	Clean around the outside surface		✓	✓	✓	✓
2	Check tightness and leakage around hydraulic connections			✓	✓	✓
3	Check electrical connections (if found)			✓	✓	✓
4	Check for offset from the zero point				✓	✓
5	Standard tests and calibration					✓

Table 3.1- BP-Measuring Instruments-02-Maintenance Scheduling

47

47

3.13- BP-Measuring Instruments-03-Installation and Maintenance

BP-Measuring Instruments-03-Installation and Maintenance:

- **General:** Avoid using adaptors.

- **Sensors:** Consider distance between sensing element and receiver.

- **Sensors:** Consider sensor placement in most representative location.

- **Pressure Gauges:** Protect P gauges from pressure surges **(See Note 1).**

- **Pressure Gauges:** Select adequate installation method **(See Note 2).**

- **Flowmeters:** Review if the flowmeter is unidirectional **(See Note 3).**

- **Flowmeters:** Avoid placing within pressure intensification **(See Note 4).**

- **Flowmeters:** Inlet and outlet must have same bore size **(See Note 5).**

48

48

Note 1: Protecting pressure gauges from pressure surges

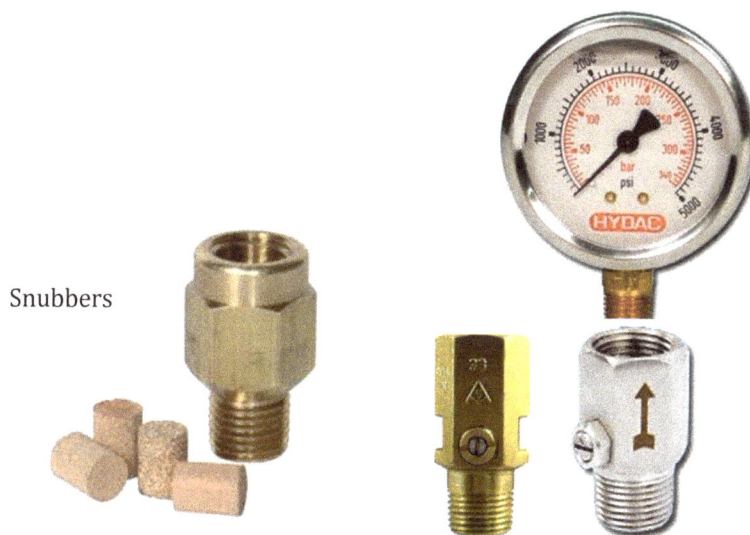

Snubbers

Fig. 3.38- Pressure Gauge Snubbers

49

49

Note 2: Select adequate installation method

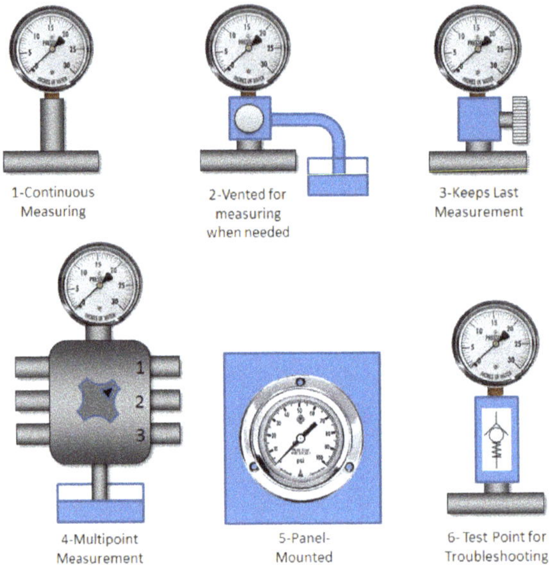

1-Continuous Measuring

2-Vented for measuring when needed

3-Keeps Last Measurement

4-Multipoint Measurement

5-Panel-Mounted

6- Test Point for Troubleshooting

Fig. 3.39- Pressure Gauge Installation

50

50

Note 3: Check if the flow meter is unidirectional

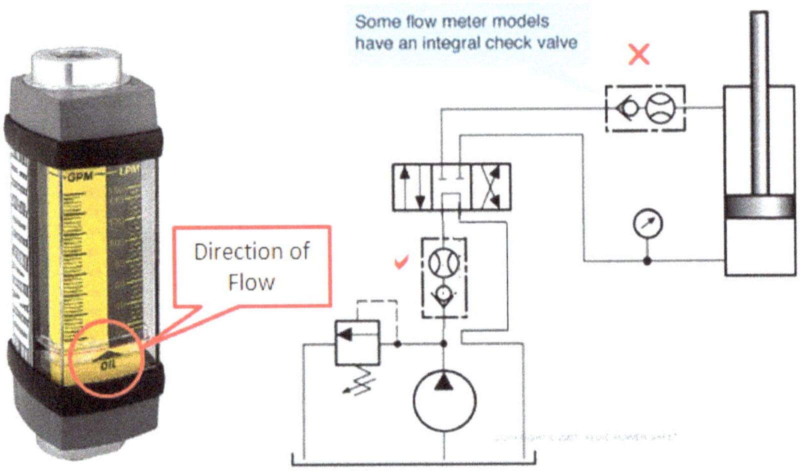

Some flow meter models have an integral check valve

Direction of Flow

Fig. 3.40- Unidirectional Flowmeters

51

51

Note 4: Avoid placing flowmeters within pressure intensification

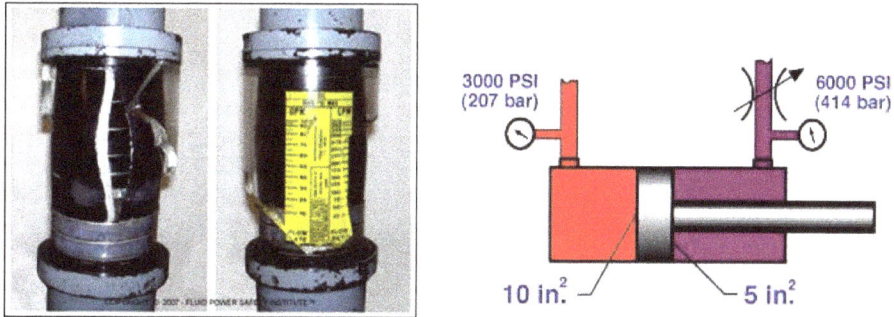

Fig. 3.41- Wrong Placement of Flowmeters
(Courtesy of Fluid Power Safety Institute)

52

52

Note 5: Size of inlet and outlet connections = size of flowmeter ports

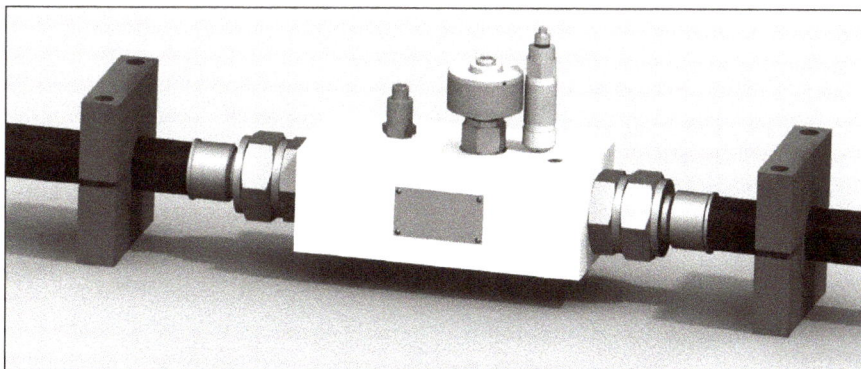

Fig. 3.42- Best Practices of Flow Meter Installation
(Courtesy of Webtec)

53

53

3.14- BP-Measuring Instruments-04-Standard Tests and Calibration

❖ Hydraulic components must be tested routinely within the scope of preventive maintenance in order to maintain reliable operation of the components/system.

❖ Accuracy of measuring instruments are affected by the following factors:

- **Repeatability:** It is the variation that can occur over the time when measuring the same parameter repeatedly.
- **Response Time:** It is the transient time the instrument takes to provide steady measurement. Response time depends on the mechanism of the device.
- **Sensitivity:** It is the minimum value to trigger the measurement only at the beginning of the scale.
- **Resolution:** It is the minimum value to trigger the measurement within the range of measurement.
- **Full Scale (Range):** It is the maximum measurement value the device can detect.

54

54

What is meant by a sensor *calibration*?   V608 (10 min)
The process of testing and adjusting the sensor when the measured value became out of an acceptable error.

What is the sensor *Error* and *Deviation*?

- Error = Measured Value – Ideal Value

- Deviation (%) = (Error/Ideal Value) x 100

What are the sources of errors in a sensor?   V609 (5 min)
One or combination of the following reasons:

- Offset from zero point.

- Mechanical wear or damage.

- Changing the scale of the operation.

55

55

BP-Measuring Instruments-04-Tests and Calibration:

- Review test and calibration instructions provided by the manufacturer.
- Test and calibration must be done by a trained person at a certified calibration center.
- Calibration and Standard Tests should be in compliance with IEC/ISO 17025.

56

56

What are the main steps of the calibration process?

- **Step 1:**
  - ○ Ideal readings from a reference device → develop the deviation curve.
  - ○ Standard points of calibration at (0, 25, 50, 75, and 100) %.

- **Step 2:**
  - ○ Check if the calibration curve below or above the allowable deviation.

- **Step3:**
  - ○ Calibration curve is above the allowable deviations →
  - ○ Perform the required adjustments to bring the device into compliance.

- **Step4:**
  - ○ Issue a certificate of calibration.

57

57

Chapter 3 Reviews

1. Measuring instruments that provide continuous measurement of a physical parameter in form of visual reading are called?
   A. Gauges/Meters.
   B. Switches.
   C. Sensors/Transducers.
   D. None of the above.

2. Measuring instruments that provide discrete "ON/OFF" electronic signal when the measured physical parameter reaches a pre-set value are called?
   A. Gauges/Meters.
   B. Switches.
   C. Sensors/Transducers.
   D. None of the above.

3. Measuring instruments that provide continuous "analog" measurement of a physical parameter in form of electronic signal are called?
   A. Gauges/Meters.
   B. Switches.
   C. Sensors/Transducers.
   D. None of the above.

4. The most accurate flow meter is?
   A. Orifice flow meter.
   B. Positive displacement flow meter.
   C. Turbine flow meter.
   D. Float flow meter.

5. One of the challenges to that should be considered when using wireless sensors is?
   A. Cyber security of the data.
   B. Accuracy of the measurements.
   C. Too much wiring.
   D. Some parameters can't be measured wirelessly.

Chapter 3 Assignment

Student Name: --- Student ID: ------------------

Date: -- Score: -----------------------

Question: List main features of traditional versus wireless sensors.

Answer:

Traditional Sensors:
- Use reliable wired sensors.
- Creates potential risk for workers in case of visual readings.
- Costs for setting up and wiring.

Wireless Sensors and IoT-Enabled Solutions:
- No risk of visual readings.
- Reduce the cost of setting and wiring.
- Remote condition monitoring from anywhere in the world.
- Smart predictive maintenance and machine diagnosis.
- Reduce unplanned downtime.
- Battery powered wireless sensors require frequent battery charge.
- Cyber security is required for the safety of a machine operation.

Chapter 4
Maintenance of Pumps

Objectives:

This chapter provides guidelines for **pumps** selection, replacement, maintenance scheduling, installation, testing, storage and transportation. This chapter is supported by examples and figures provided by leading fluid power manufacturers.

Brief Contents:

4.1-BP-Pumps-01-Selection and Replacement
4.2-BP-Pumps-02-Maintenance Scheduling
4.3-BP-Pumps-03-Installation and Maintenance
4.4-BP-Pumps-04-Standard Tests and Calibration
4.5-BP-Pumps-05-Transportation and Storage

0

0

4.1- BP-Pumps-01-Selection and Replacement
4.1.1- Selecting or Replacing Pumps

When selecting or replacing an existed pump →

BP-Pumps-01-Selection and Replacement:
- Review maximum/optimum operating pressure.
- Review min/maximum/optimum operating speed.
- Review maximum overall efficiency at the optimum operating conditions.
- Review size (**See Note 1**) and displacement control requirements.
- Review type of fluid.
- Review contamination tolerance.
- Review noise level.
- Review initial cost.
- Review approximate service life.
- Review availability and interchangeability.
- Review maintenance and spare parts.
- Review physical size and weight.

1

1

Note 1:

If a pump is oversized, one or more of the following could occur:

- Increased actuators speed → improper sequence & unsafe operation.

- Increased flow forces on spool valves → improper valve operation.

- Turbulent flow → all relevant consequences.

- pressure drops across components ↑ → wasted energy ↑ → heat generation ↑.

If a pump is undersized, one or more of the following could occur:

- Actuators slowdown → improper sequence & unsafe operation → less productive machine.

- Large valves lose controllability for small flow. For example, if the valve was originally a pilot operated valve, it will work improperly for low inlet flow.

2

2

4.1.2- Displacement Calculation of Legacy Pumps

This information is helpful when replacing a legacy pump

External Gear Pumps:

$$\textbf{Pump Displacement} = \textbf{L} \times \textbf{W} \times \textbf{H} \times \textbf{N} \times 2 = \textbf{L} \times \textbf{W} \times \frac{[\textbf{D}_1 - \textbf{D}_2]}{2} \times \textbf{N} \times 2 \qquad 4.1$$

$\textbf{N}$ = No. of teeth of one gear

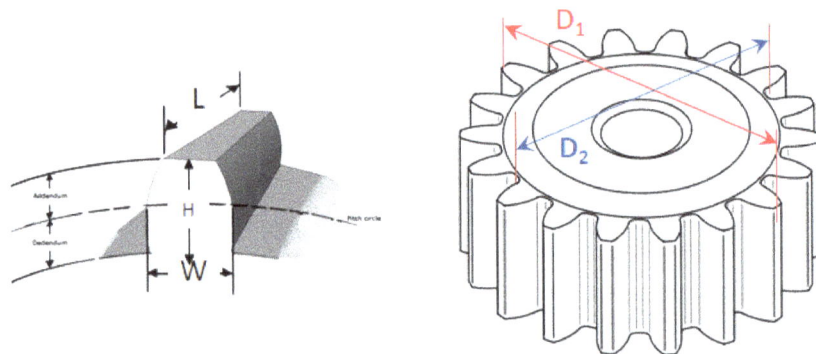

Fig. 4.1- Measurements for Calculating Size of External Gear Pump

3

3

Internal Gear Pumps:

Pump Displacement $= N \times L \times [A_{max} - A_{min}]$ **4.2**

N = No. of inner gear teeth and L = width of the gear teeth.

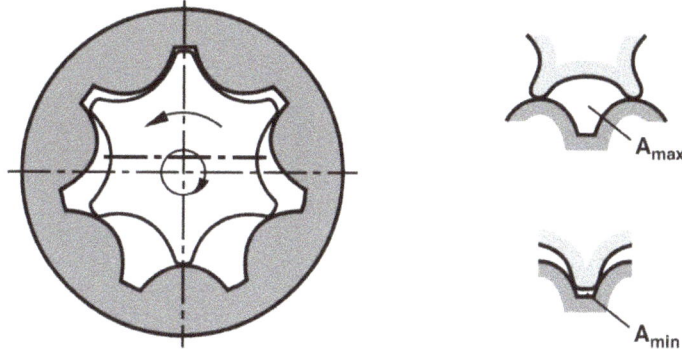

Fig. 4.2- Measurements for Calculating Size of Gerotor
(Courtesy of Bosch Rexroth)

4

4

Screw Pump:

Pump D isplacement $= \dfrac{\pi}{4} \times [D^2 - d^2] \times s \times c$ **4.3**

c = correction factor < 1 due to the interlocking of threads of both spindles.

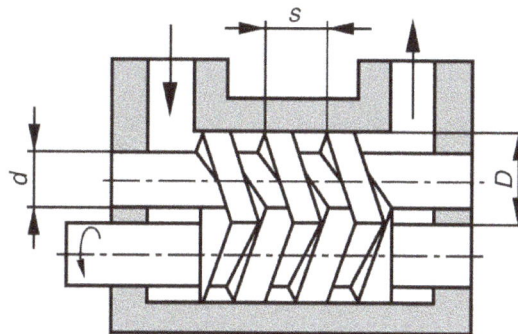

Fig. 4.3- Measurements for Calculating Size of Screw Pump
(Courtesy of Bosch Rexroth)

5

5

Unbalanced Vane Pumps:

$$\text{Pump Displacement} = 2 \times \pi \times b \times e \times D \qquad 4.4$$

Balanced Vane Pumps:

$$\text{Pump Displacement} = \frac{\pi}{4} \times [D^2 - d^2] \times b \times 2 \qquad 4.5$$

Where **b** = Vane width and **e** = eccentricity.

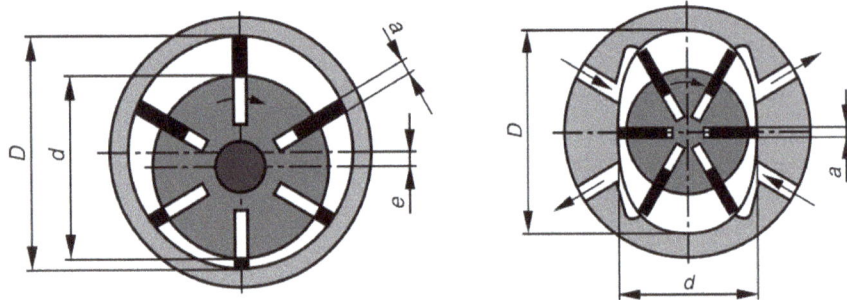

Fig. 4.4- Measurements for Calculating Size of Vane Pump
(Courtesy of Bosch Rexroth)

6

6

Radial Piston Pumps:

$$\text{Pump Displacement} = 2 \times e \times z \times \left[\frac{\pi \times d_k^2}{4}\right] \qquad 4.6$$

Where d_k = piston diameter, **e** = eccentricity, and **z** = number of pistons.

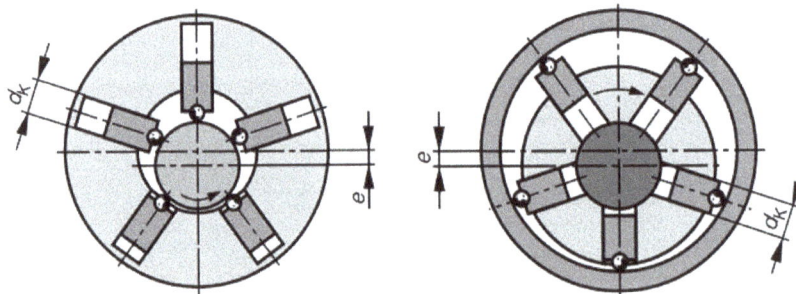

Fig. 4.5- Measurements for Calculating Size of Radial Piston Pump
(Courtesy of Bosch Rexroth)

7

7

Swash Plate Pump:

$$\text{Pump Displacement} = 2 \times r_h \times z \times \sin \alpha \left[\frac{\pi \times d_k^2}{4} \right] \qquad 4.7$$

d_k = piston diameter,
α = angle of inclination,
z = number of pistons.

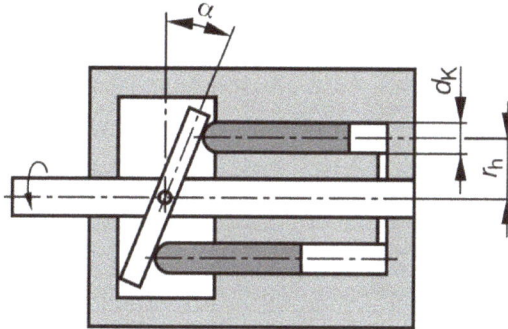

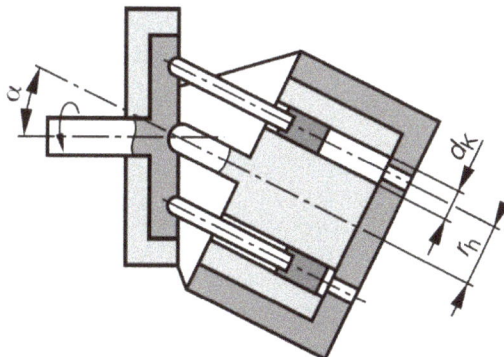

Fig. 4.6- Measurements for Calculating Size of Swash Plate Pump
(Courtesy of Bosch Rexroth)

8

8

Bent Axis Pumps:

$$\text{Pump Displacement} = 2 \times r_h \times z \times \tan \alpha \left[\frac{\pi \times d_k^2}{4} \right] \qquad 4.8$$

d_k = piston diameter,
α = angle of inclination,
z = number of pistons.

Fig. 4.7- Measurements for Calculating Size of Bent Axis Pump
(Courtesy of Bosch Rexroth)

9

9

4.2- BP-Pumps-02-Maintenance Scheduling

Unless otherwise stated by components and systems manufacturer:

#	Preventive Maintenance Actions	Daily	Weekly	Monthly	Biannually	Annually
1	Clean around and outside surface	✔	✔	✔	✔	✔
2	Check for unusual sound	✔	✔	✔	✔	✔
3	Check temperature of the pump body	✔	✔	✔	✔	✔
4	Check tightness and leakage around hydraulic connections		✔	✔	✔	✔
5	Check electrical connections (if found)		✔	✔	✔	✔
6	Check for vacuum readings (if found)			✔	✔	✔
7	Check for vibration and condition of dampers			✔	✔	✔
8	Standard tests and calibration					✔

Table 4.1- BP-Pumps-02-Maintenance Scheduling

10

10

SUGGESTED MAINTENANCE ON HYDRAULIC POWER UNITS
(Courtesy of Womack)

❖ **Daily Inspections:**

▪ **Inspect Visually for Oil leakage:**
  o Around shafts of pumps, cylinders, and hydraulic motors.
  o Tighten any fittings which leak.

▪ **Inspect for Oil Level:**
  o Observe oil level in the reservoir over a complete cycle of the machine.
  o Make sure that oil level never falls below the LOW mark in any phase of the machine operation.

11

11

❖ **Weekly Inspections:**
- **Housekeeping inspection:**
  - Clean up spilled oil. Clean dirt off top of reservoir.

- **Inspect Breather:**
  - Make sure filler and breather caps are in place.
  - Check condition of breather cap.
  - Remove accumulations of lint and dirt.
  - Breather must be cleaned, especially at higher altitudes where air pressure is lower.

- **Inspect Fluid and Water Content:**
  - Obtain a small volume of fluid from a drain line.
  - Visually inspect color, odor, and viscosity, and water content.
  - If water is found:
    - ➤ Check water leakage from a shell and tube heat exchanger.
    - ➤ Check water condensation due to atmospheric temperature changes around the reservoir. If this is the case, it will be necessary to tap off accumulated water every day. 12

12

❖ **500-Hour Intervals:**
- **Filter Element:** Filter replacement and strainer cleaning.

- **Hydraulic Oil:** Make sure to keep 10 to 25% of reservoir capacity as a reserve supply.

- **Hoses:** All hoses should be replaced after being in service approximately 5 years even though they appear to be in good condition.

- **Cylinders Overhaul:** Overhaul kits can be purchased from cylinder manufacturer and used within 2 years. Rubber deteriorates by oxidation in a few years.

- **Pumps Overhaul:** Overhaul kits include (soft seals, gaskets, bearing, and pump hard parts). In some important machine, an entire pump could be kept as spare.

- **Electrical Parts:** Keep spare coils for solenoid valves, relays, and motor starters. Limit switches, fuses, lamp bulbs, and on important machines, spare relays should be kept.

13

13

4.3- BP-Pumps-03-Installation and Maintenance

1. Install the Pump to avoid Cavitation.
2. Identify Ports and Direction of Rotation.
3. Proper Placement with the Reservoir.
4. Proper Mounting for Direct Drive.
5. Proper Coupling for Direct Drive.
6. Proper Shaft Alignment for Direct Drive.
7. Proper Indirect (Side) Drive.
8. Adequately Damp Vibration.
9. Proper Oil Intake and Return.
10. Proper Priming.
11. Proper Case Drain.
12. Proper Oil Discharge.
13. Review Range of Driving Speed.
14. Review Range of Working Pressure.
15. Review Range of Working Temperature.
16. Review Compatibility with Working Hydraulic Fluid.
17. Review Prime Mover Overloading Conditions.
18. Proper Placement of Hydraulic Power Unit.
19. Proper Installation of Hydrostatic Transmission

14

14

4.3.1- Install the Pump to Avoid Cavitation

Every effort must be made to avoid developing cavitation.

The following list shows actions that cause pump cavitation:

- Incorrect pump ports identification
- Incorrect pump direction of rotation.
- Improper pump placement with the reservoir.
- Improper pump oil Intake.
- Improper pump priming.
- Increased pump driving speed.
- Increased pump inlet temperature.
- Incorrect oil viscosity.

Video 281 (4 min)

15

4.3.2- Identify Ports and Direction of Rotation

- Suction and Discharge ports are relevant to its direction of rotation.
- If a pump is connected or rotates oppositely → pump fail.

How to Identify Pump Ports:

1. knowing how a pump works.

2. A symbol on the name plate matches the physical location of ports.

3. Suction port is larger than discharge port in unidirectional pumps.

4. Unbalanced pumps have bearing wear signs at the discharge side.

16

16

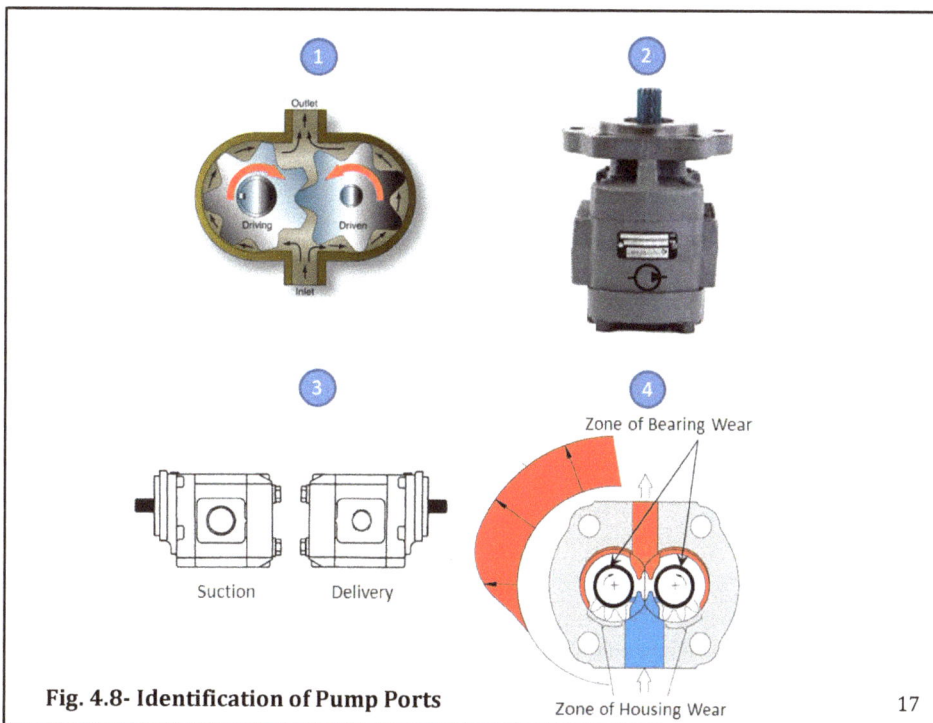

Fig. 4.8- Identification of Pump Ports

17

How to Identify Pump Direction of Rotation:

- Pumps could be unidirectional, bidirectional or over-center type.

- Over-center and unidirectional pumps have a predefined direction of rotation (CW or CCW).

- Signs that reveals the direction of rotation:

  1. knowing how a pump works and location of pump ports.

  2. Visual symbol on the pump housing.

  3. Review the ordering code or the name plate of the pump (**See Note 1**).

18

18

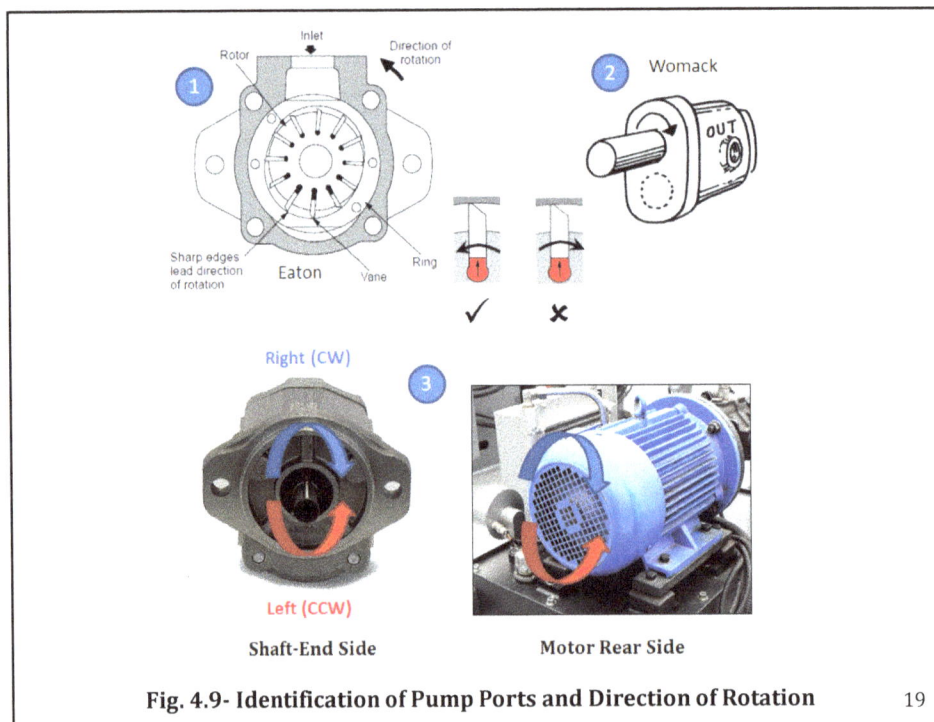

Fig. 4.9- Identification of Pump Ports and Direction of Rotation   19

19

Switching a Pump Direction of Rotation:

For a Bidirectional Pumps:

1. When the prime over is an electric motor.
2. When the prime mover is an engine.

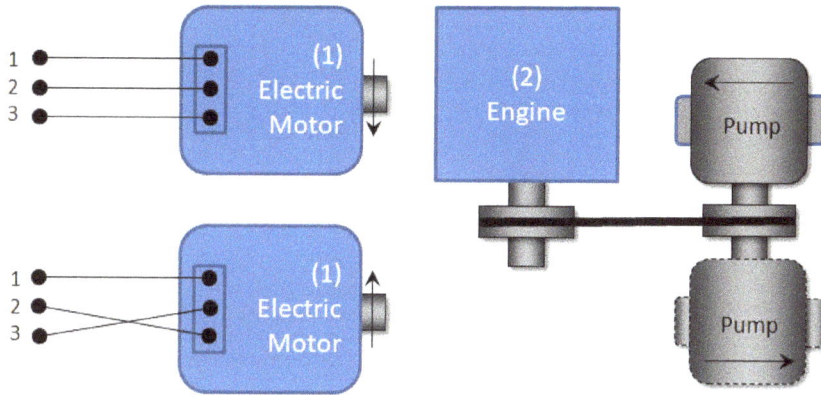

Fig. 4.10- Switching the Direction of Rotation

20

20

For a Unidirectional Pumps:

Some of pump designs require disassembling and reconfiguring the pump as per the pump manufacturer before switching them to run oppositely.

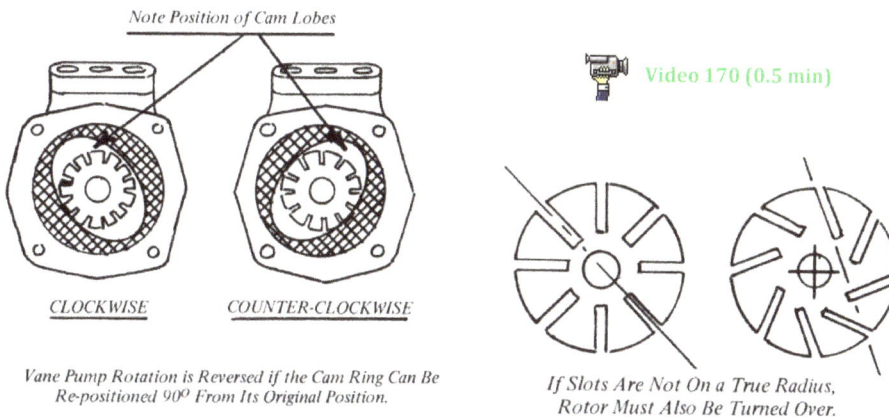

Note Position of Cam Lobes

Video 170 (0.5 min)

CLOCKWISE   COUNTER-CLOCKWISE

Vane Pump Rotation is Reversed if the Cam Ring Can Be Re-positioned 90º From Its Original Position.

If Slots Are Not On a True Radius, Rotor Must Also Be Turned Over.

Fig. 4.11- Instructions for Switching A Vane Pump Direction of Rotation (Courtesy of Womack)

21

21

4.3.3- Proper Placement with the Reservoir

Common scenarios of placing a pump with respect to the reservoir:

1. Small pumps.

2. Medium size pumps.

3. Larger size pumps.

4. Not as common (a pump is immersed + better cooling + difficult to monitor the pump condition).

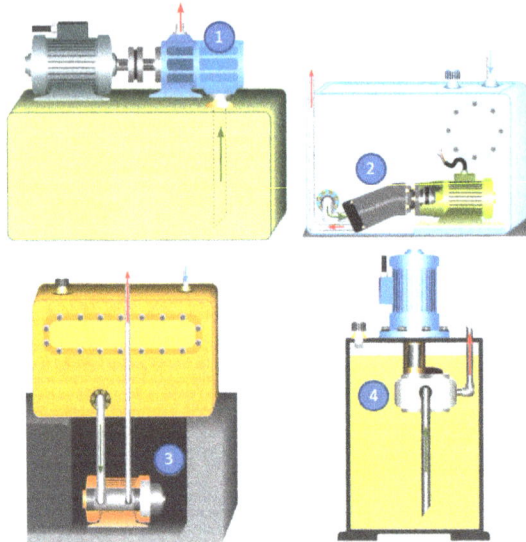

Fig. 4.12- Proper Placement with the Reservoir (Courtesy of Assofluid)

22

22

4.3.4- Proper Mounting for Direct Drive.

- **For Safe Operation:** Rotating parts should be covered and guarded.

- **Industrial Applications:** Direct Drive is common where a pump is directly driven by an electric motor.

- **Mobile Application:** A main pump is directly driven by an engine using a special mount and couplings.

23

23

Foot-Mounting:

- A traditional method.
- Pump and motor are mounted independently on two different frames.
- **Supporting Base:** if not rigid enough → pump-motor misalignment.

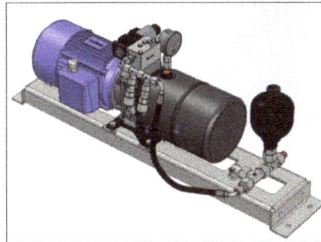

Fig. 4.13- Foot-Mounting for Direct Drive

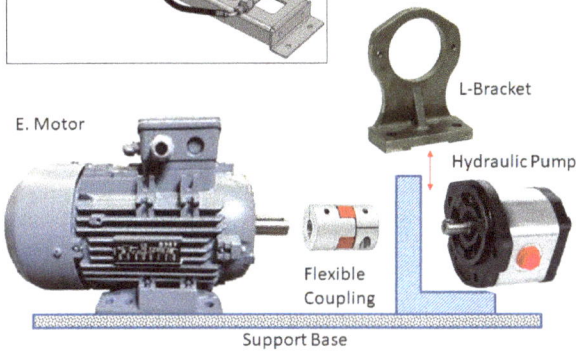

24

24

- **L-Bracket:** L-bracket is an adaptor used to align the pump-motor shafts.
- **Shaft Alignment:**
  o Shaft alignment is challenging.
  o Must be redone every time the pump is removed from the bracket.

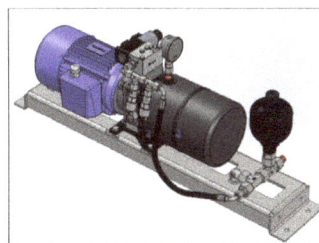

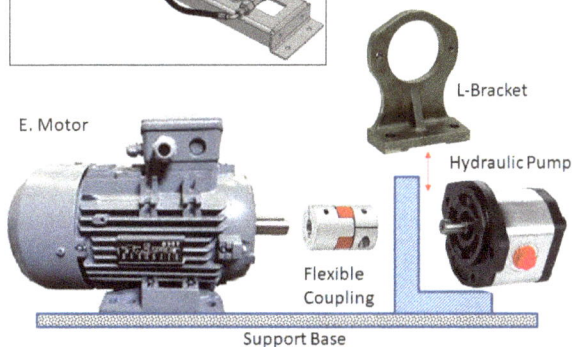

25

25

- **Coupling:**
  - ○ Flexible couplings are recommended → absorb vibration.
- **Distance between Flanges:**
  - ○ It is flexible to adapt to the exact required distance.

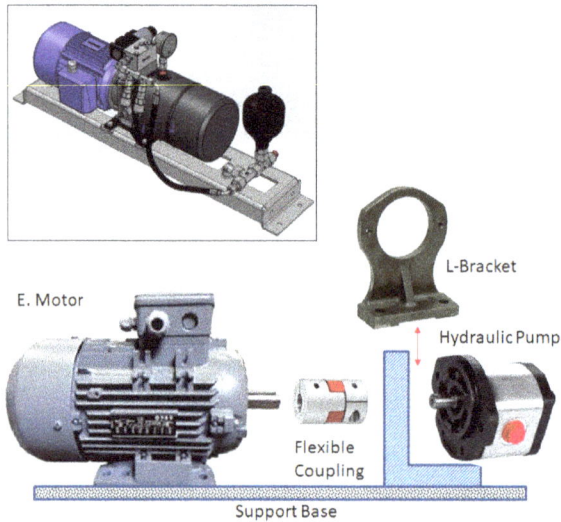

26

Bellhousing Mounting:

- A common method (Bell housing) → easy and a quick assembly
- Pump and motor are mounted on one frame of reference.

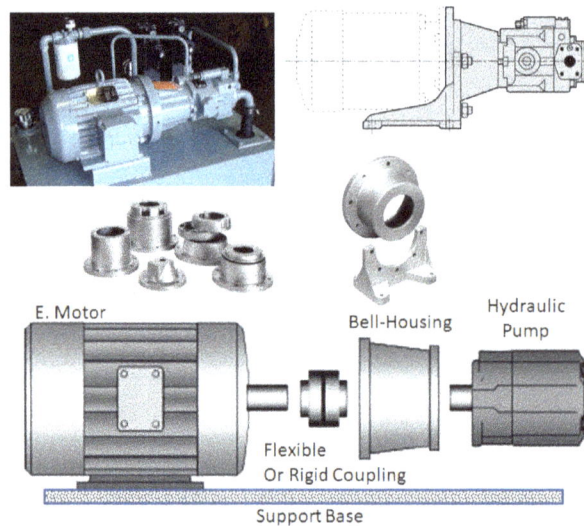

Fig. 4.14- Bellhousing Mounting for Direct Drive

27

- **Supporting Base:**
  - ❑ A large electric motor is foot-mounted to the supporting base.
  - ❑ The motor flange supports the weight of the bell housing and the pump.
  - ❖ For small electric motors, a bell housing is foot-mounted to the base.
  - ❖ The bell housing supports the weight of both the motor and the pump.

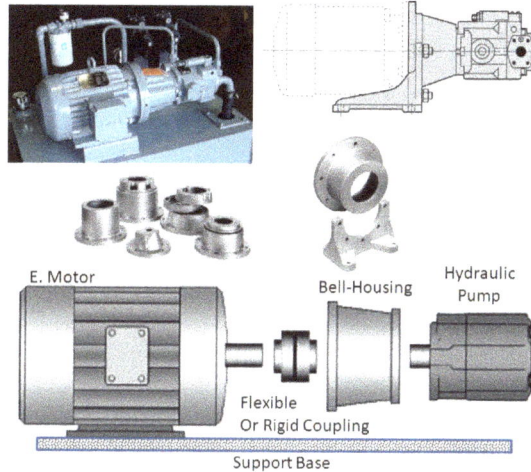

28

28

- **Bellhousing:**
  - o Bell-shaped housing is designed to integrate the pump and the motor.
  - o Available in standard lengths to meet various pump-motor combinations.

- **Shaft Alignment:**
  - o Is considered in the design of the Bell Housing.
  - o No need to redo the shaft alignment if the ump is removed.

- **Coupling:**
  - o Flexible or rigid couplings are used.
  - o However, flexible couplings are still recommended.

- **Distance between Flanges:**
  - o It is fixed and must be considered in the Bellhousing design.

29

29

4.3.5- Proper Coupling for Direct Drive

Best practices for installing a coupling:

- Be capable of continuously withstanding the maximum torque.

- Check for perfect fit on the pump and motor shafts.

- Loose fitting couplings → accelerated wear of the drive shaft.

- DO NOT push hardly the coupling onto the pump drive shaft. If it is too tight, it may be necessary to heat coupling for installation.

- Drive couplings shall be provided with a suitable protective guard when the coupling area is accessible during operation of the pump.

30

30

Rigid Coupling:

- Common for Belhousing mounting.

- Two metallic symmetric parts.

- Sized for standard shafts.

- Do not tolerate shaft misalignment

- Do not tolerate shaft end backlashes.

- Compact design.

- Low weight and mass moment of inertia.

- Independent of direction of rotation (suitable for reversing operation).

Rigid Coupling

Flexible Coupling

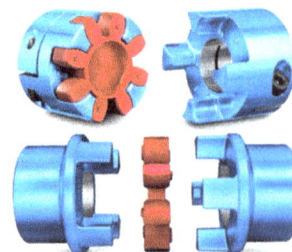

Fig. 4.15- Coupling Configurations for Direct Drive (www.flender.com)

31

31

Flexible Coupling:

Rigid Coupling

- Common for Foot mounting.

- Two metallic symmetric parts.

- Sized for standard shafts.

- A rubber-based flexible element is placed between the two halves of the coupling

- Tolerate shaft misalignment

- Damp vibration.

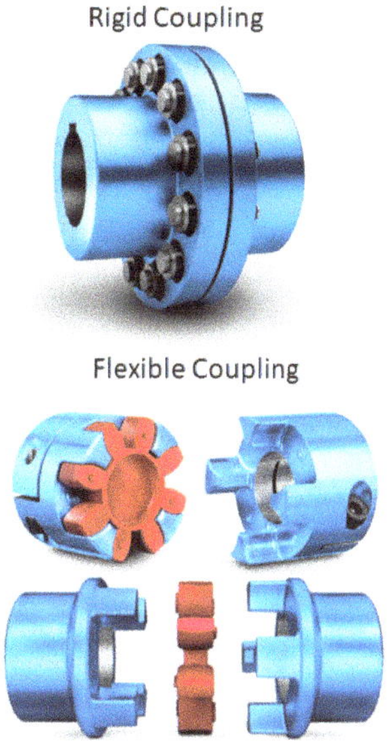

Flexible Coupling

Video 171 (0.5 min)   Video 127 (2.5 minutes)

32

32

4.3.6- Proper Shaft Alignment for Direct Drive

❖ Ideally, pump-motor shafts should be coaxial and aligned to each other.
❖ Misalignment:
- Axial misalignment (known also as Offset).
- Angular misalignment (known also as Angularity).

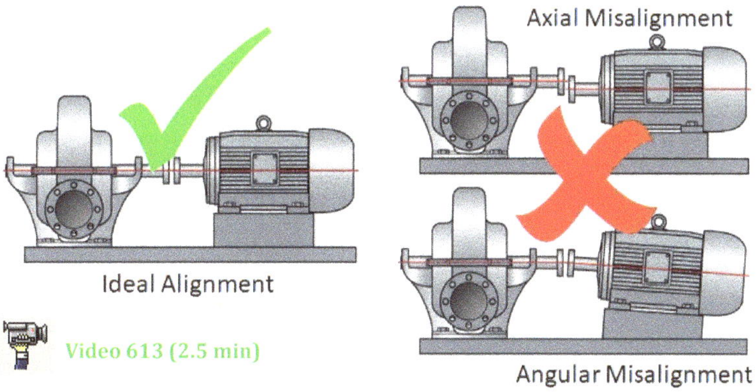

Axial Misalignment

Ideal Alignment

Video 613 (2.5 min)

Angular Misalignment

Fig. 4.16- Axial and Angular Shaft Misalignment (wastewater101.net)

33

33

❖ Shaft misalignment →:

- Increased vibration.

- Damage to bearings and couplings.

- Shaft seal failure.

- Wear of pump internal parts.

- Excessive heat and energy loss.

Video 172 (0.5 min)

Video 614 (2 min)

❖ Therefore, pump-motor shaft alignment should be checked:

- During pump installation.

- Routinely during major maintenance.

- When there is any sign of increased noise and vibration.

34

34

Dial Indicator:

- Is a tool used for checking shaft misalignment.
- A precise gear mechanism that is driven by a plunger.
- A very fine measuring resolution (0.001" = 1.0 mil).
- Available with digital readings.
- Both should have means to adjust the zero point of the indicator.

Video 611 (3 min)   **Fig. 4.17- Dial Indicators**

35

35

- Permissible misalignment is inversely proportional to the rotational speed.
- Axial misalignment (offset) is measured in mils.
- Angular misalignment (slope) is measured in mils/inch.

Permissible Misalignment				
	Angular Misalignment (Mils)		Axial Misalignment (Mils/Inch)	
RPM	Excellent	Acceptable	Excellent	Acceptable
3600	0.3/1"	0.5/1"	1.0	2.0
1800	0.5/1"	0.7/1"	2.0	4.0
1200	0.7/1"	1.0/1"	3.0	6.0
900	1.0/1"	1.5/1"	4.0	8.0

Table 4.2- Example of Pump Shaft Permissible Misalignment

Video 610 (3 min)

Best practices for measuring misalignment:
- Rotate the pump shaft in the assigned direction of the pump rotation.
- Offset and angularity are measured on horizontal and vertical planes.
- However, they can be checked on any plane.

36

36

Checking Axial Misalignment (Offset):
- Place the dial indicator on the rim.
- Make sure the plunger is in the middle of its stroke.
- Then bring the zero pint on the scale at the place of the dial.

Vertical Offset:
- Start at 12 o'clock and turn the pump shaft 180 degrees till 6 o'clock.
- Vertical Total Indicator Reading (TIR) = Reading from top to bottom.
- Actual Vertical Misalignment $Y = TIR/2$ measured in mils.

Horizontal Offset:
- Start at 3 o'clock and turn the pump shaft 180 degrees till 9 o'clock.
- Horizontal Total Indicator Reading (TIR) = Reading from side to side.
- Actual Horizonal Misalignment $X = TIR/2$ measured in mils.

37

37

Checking Angular Misalignment (Angularity):
- Place the dial indicator on the face.
- Make sure the plunger is in the middle of its stroke.
- Then bring the zero pint on the scale at the place of the dial.

Vertical Angularity:
- Start at 12 o'clock and turn the pump shaft 180 degrees till 6 o'clock.
- Vertical Total Indicator Reading (TIR) = Reading from top to bottom.
- Actual Vertical Angularity = TIR/Coupling Diameter.

Horizontal Angularity:
- Start at 3 o'clock and turn the pump shaft 180 degrees till 9 o'clock.
- Vertical Total Indicator Reading (TIR) = Reading from side to side.
- Actual Vertical Angularity = TIR/Coupling Diameter (D).

Video 612 (5 min)

38

38

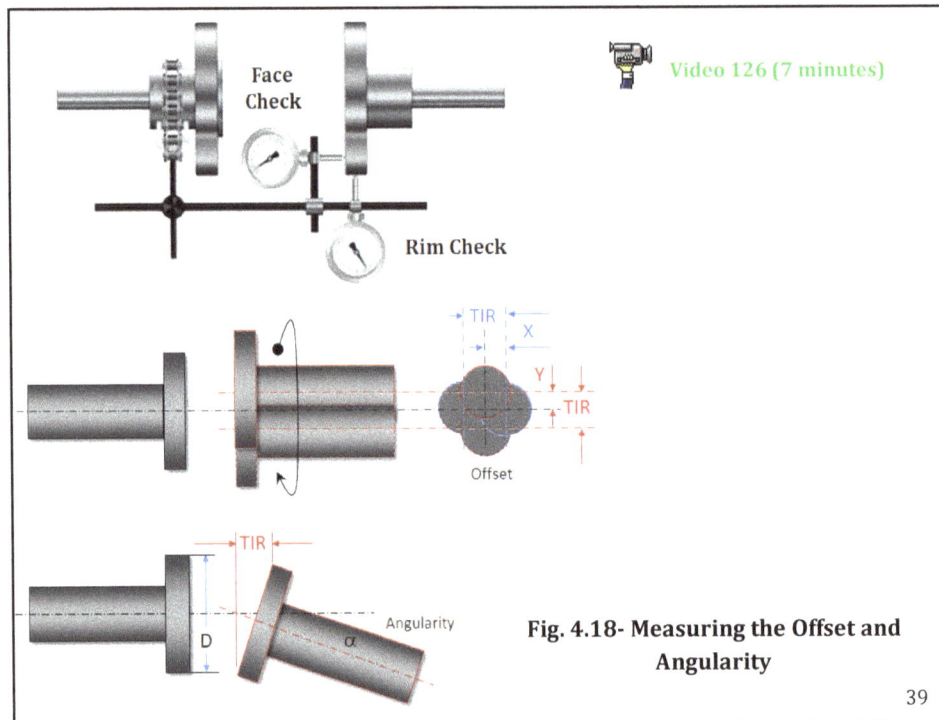

Video 126 (7 minutes)

Face Check

Rim Check

Fig. 4.18- Measuring the Offset and Angularity

39

39

4.3.7- Proper Indirect (Side) Drive

- Side Drive through an intermediate element such as belt, gear, chains, etc.
- Common in mobile applications where an engine is a prime mover.
- Occasionally in industrial applications when an electric motor drives multiple pumas.

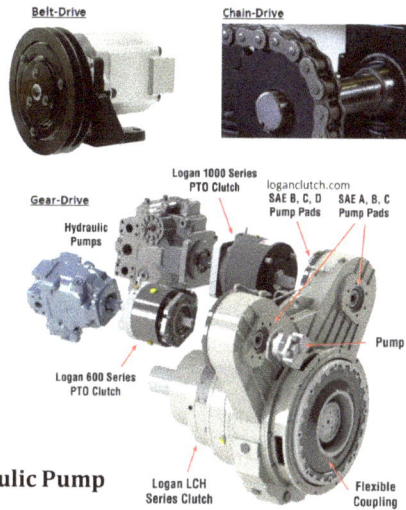

Fig. 4.19- Side Drive of a Hydraulic Pump

40

40

- Side drives → side load on shaft bearings.
- Exceeding permissible side load → shaft misalignment.

Axial Piston Fixed Pump A2FO

Technical data

Permissible radial and axial forces of the drive shafts

(splined shaft and parallel keyed shaft)

Size	NG		5	5[3)]	10	10	12	12	16	23	23
Drive shaft	ø	mm	12	12	20	25	20	25	25	25	30
Maximum radial force[1)] at distance a (from shaft collar)	$F_{q\,max}$	kN	1.6	1.6	3.0	3.2	3.0	3.2	3.2	5.7	5.4
	a	mm	12	12	16	16	16	16	16	16	16
with permissible torque	T_{max}	Nm	24.7	24.7	66	66	76	76	102	146	146
≙ permissible pressure Δp	Δp_{perm}	bar	315	315	400	400	400	400	400	400	400
Maximum axial force[2)]	$+F_{ax\,max}$	N	180	180	320	320	320	320	320	500	500
	$-F_{ax\,max}$	N	0	0	0	0	0	0	0	0	0
Permissible axial force per bar operating pressure	$\pm F_{ax\,perm/bar}$	N/bar	1.5	1.5	3.0	3.0	3.0	3.0	3.0	5.2	5.2

**Table 4.3- Example of Pump Shaft Permissible Side Loads
(Courtesy of Bosch Rexroth)**

41

41

Best Practices for Pump Side Drives:

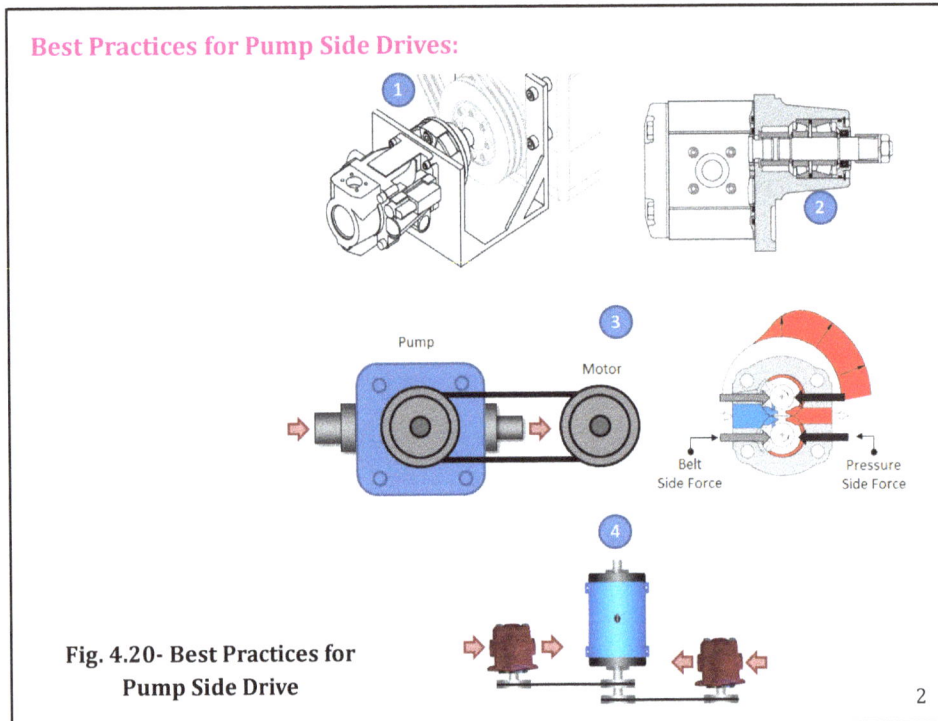

Fig. 4.20- Best Practices for
Pump Side Drive

42

4.3.8- Adequately Dampen Vibration

- Positive displacement pump → flow pulsation → Pressure ripples →
- Vibration and noise control is out of scope of this textbook.
- However, vibration dampers → absorb vibration.

Fig. 4.21- Conditions of Vibration Dampers

43

43

- Flanges, clamp bolts, support surface must be covered with antivibration elements.

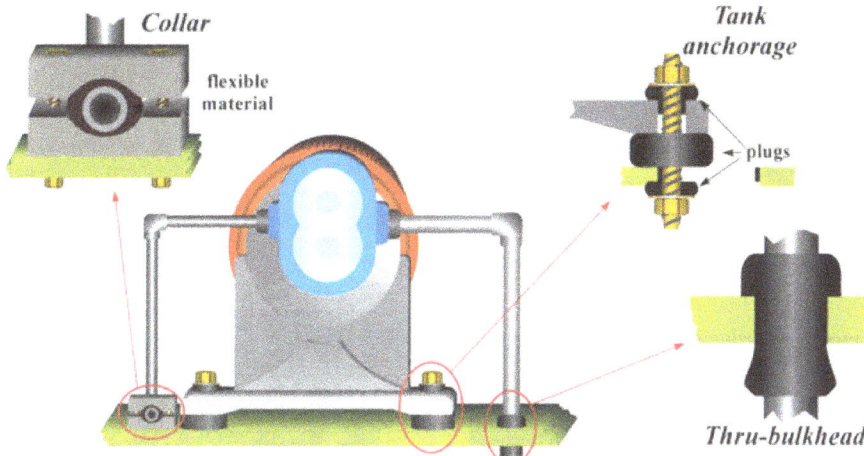

Fig. 4.22- Tools for Vibration Damping (Courtesy of Assofluid)

44

44

4.3.9- Proper Oil Intake and Return

- ❖ **Line Diameter:** It is predefined during design stage based on:
- Pump port size + Avoid turbulence + moderate flow speed + minimize pressure losses + avoid cavitation.
- ❖ **Line (placement + bottom clearance + length).**

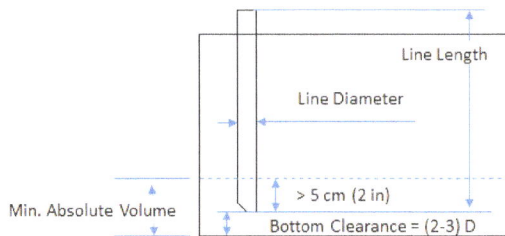

Fig. 4.23- Pump Suction Line Size and Placement

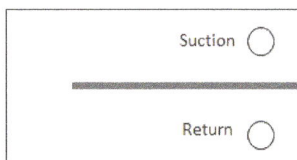

With Baffle Plate

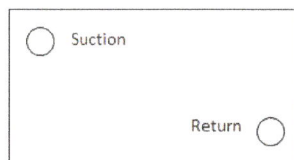

With no Baffle Plate

45

45

❖ **Line Routing:**
- Improper suction line routing → possible cavitation.
- Improper return line routing → possible backpressure.

❖ **Line Type:**
- DO NOT replace hard tubing suction lines by flexible hoses.
- DO NOT use pressure hose in suction lines.

❖ **Suction Strainer:** Must be routinely must be cleaned.

Fig. 4.24- Result of Suction Strainer Replacement Time Overdue

46

46

4.3.10- Proper Priming
- **Purpose**: Priming → bleed air from inside the mechanism →
- avoid air lock and dry run.
- **Priming Fluid:** must be same as the fluid used in the system.
- **Identify Filling/Bleed Holes based on Pump Orientation:**

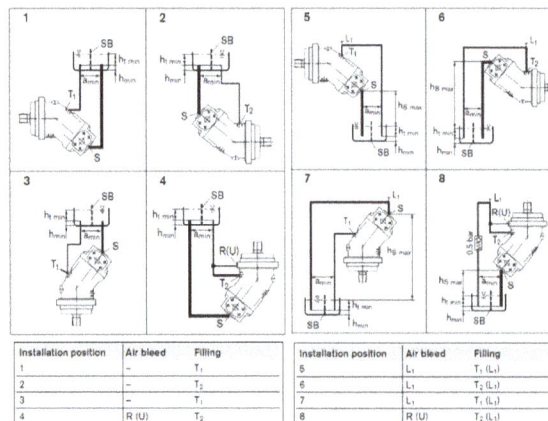

Installation position	Air bleed	Filling
1	–	T_1
2	–	T_2
3	–	T_1
4	R (U)	T_2

Installation position	Air bleed	Filling
5	L_1	$T_1 (L_1)$
6	L_1	$T_2 (L_1)$
7	L_1	$T_1 (L_1)$
8	R (U)	$T_2 (L_1)$

**Fig. 4.25- Example for Identifying the filling/Bleed Ports
(Courtesy of Bosch Rexroth)**

47

47

❖ **Priming Procedure (Fig. 4.26):** Video 174 (0.5 min)

- Route the pump outlet to the tank.
- Ensure the filling devices are clean to manufacturer's recommendation.
- Identify the filing port.
- Fill the housing with clean priming fluid via the identified filling port.
- Rotate the pump shaft till it displaces bubble-free oil from outlet.
- Close the pump priming hole and assemble the pump to prim mover.
- Run the pump and check if it displaces bubble-free oil for 20 seconds.

Pour Oil Into Pump Case.

Rotate Pump by Hand With Fitting Cracked on Outlet Line.

Fig. 4.26- Hydraulic Pump Priming Procedure (Courtesy of Womack) 48

48

4.3.11- Proper Case Drain

- Route the case drain line so that it remains full of fluid (non-siphoning).
- Case drain pressure must be < manufacturer's maximum value.
- If no information found ≈ 1.7 bar (25 psi).
- Pumps with more than one case drain port → use the highest port.
- Case drain line must be a separate, unrestricted, full sized and connected directly to the reservoir below the lowest fluid level.
- If a filter is placed on the case drain, make sure is sized and cleaned properly.

4.3.12- Proper Oil Discharge

- Avoid developing turbulent flow by proper sizing
- Recommended flow speed 2.1-4.6 m/s (7-15 foot/s).
- Select the discharge line to meet the maximum system pressure.

49

4.3.13- Review Range of Driving Speed

Overspeed:

- Mechanical efficiency ↓
- Possible cavitation.
- Possible failure due friction.
- Reduced bearing lifetime.

Under-speed:

- Volumetric efficiency ↓
- Increased leakage and heats.
- Vane and piston pump ability to work under high pressure.

Optimum Speed:

- Defined by manufacturer.
- At the maximum overall efficiency.

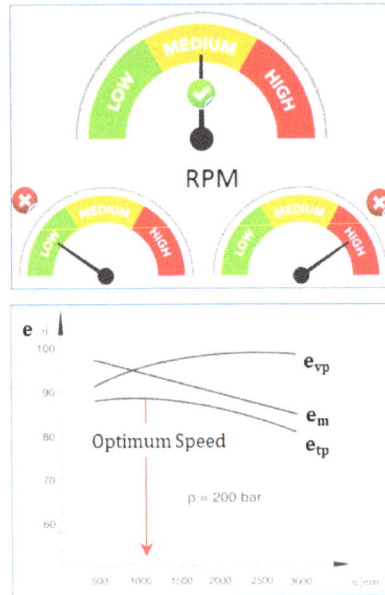

Fig. 4.27- Recommended and Optimum Driving Speeds of a Pump

50

50

4.3.14- Review Range of Working Pressure

Overpressure:

- Volumetric efficiency ↓
- Increased leakage and heats.
- Possible mechanical failure.
- Reduced pump bearing life.

Pressure Spikes:

- Fatigue failure.

Low Working Pressure:

- Mechanical efficiency ↓

Optimum Pressure:

- Defined by manufacturer.
- At the maximum overall efficiency

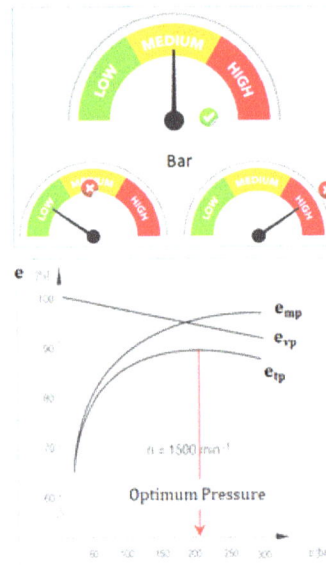

Fig. 4.28- Recommended and Optimum Working Pressure of a Pump   51

51

4.3.15- Review Range of Working Temperature

High Temperature:

- Reduced fluid viscosity → increased leakage → lack of lubrication.
- Reduced sealing element performance
- Fluid degradation → varnish formation.

Low Temperature:

- Cold start → Pump cavitation.

Ideal Temperature:

- Predefined by pump manufacturers.
- If no information is found, should not exceed (40-60) $^{\circ}$C.

52

Viscosity and temperature of hydraulic fluid

	Viscosity [mm²/s]	Temperature	Comment
Transport and storage at ambient temperature		$T_{min} \geq$ -50 °C $T_{opt} =$ +5 °C to +20 °C	factory preservation: up to 12 months with standard, up to 24 months with long-term
(Cold) start-up[1]	$v_{max} =$ 1600	$T_{St} \geq$ -40 °C	t ≤ 3 min, without load (p ≤ 50 bar), n ≤ 1000 rpm (for sizes 5 to 200), n ≤ 0.25 • n_{nom} (for sizes 250 to 1000)
Permissible temperature difference		$\Delta T \leq$ 25 K	between axial piston unit and hydraulic fluid
Warm-up phase	v < 1600 to 400	T = -40 °C to -25 °C	at p ≤ 0.7 • p_{nom}, n ≤ 0.5 • n_{nom} and t ≤ 15 min
Operating phase			
Temperature difference		$\Delta T =$ approx. 12 K	between hydraulic fluid in the bearing and at port T.
Maximum temperature		115 °C	in the bearing
		103 °C	measured at port T
Continuous operation	v = 400 to 10 $v_{opt} =$ 36 to 16	T = -25 °C to +90 °C	measured at port T, no restriction within the permissible data
Short-term operation[2]	$v_{min} \geq$ 7	$T_{max} =$ +103 °C	measured at port T, t < 3 min, p < 0.3 • p_{nom}
FKM shaft seal[1]		T ≤ +115 °C	see page 5

1) At temperatures below -25 °C, an NBR shaft seal is required (permissible temperature range: -40 °C to +90 °C).
2) Sizes 250 to 1000, please contact us.

Fig. 4.29- Example of Manufacturer's Instructions about Working Temperature for A2FO Pump (Courtesy of Bosch Rexroth)

53

4.3.16- Review Compatibility with working Hydraulic Fluid

❖ Positive displacement pumps/motors are designed for use of hydraulic fluids that can lubricate the bearings and internal parts.

❖ They should never be used on fluids such as water, kerosene, fuel oil, jet fuel, gasoline, etc. because these fluids do not provide the necessary lubrication.

❖ Consult pump/motor manufacturers provide for compatible fluids.

❖ If no information were found, these general rules may be followed:

❖ **Petroleum-Based (Mineral) Fluid for Industrial Hydraulic Systems:**
▪ Used for most power units
▪ Compounded with the proper additives.
▪ Recommended viscosity is (22 – 100) cSt, i.e. (100 – 500) SSU
▪ At a 37°C (100° F).

54

54

❖ **Multi-Grade Fluids for Mobile Hydraulic Systems:**

▪ Used for applications that required high Viscosity Index.

▪ Operate satisfactorily over a wide range of ambient temperatures.

▪ Example: automatic transmission.

❖ **Fire-Resistant Fluids for High Working Temperature:**

▪ Used for applications with hazard of fire.

▪ Example: die casting and steel mills.

❖ **Synthetic Fluids for Special Applications:**

▪ Used for applications that require special fluid requirements.

▪ Example: Aerospace.

55

55

Technical data

Hydraulic fluid

Before starting project planning, please refer to our data sheets RE 90220 (mineral oil), RE 90221 (environmentally acceptable hydraulic fluids), RE 90222 (HFD hydraulic fluids) and RE 90223 (HFA, HFB, HFC hydraulic fluids) for detailed information regarding the choice of hydraulic fluid and application conditions.

The fixed pump A2FO is not suitable for operation with HFA hydraulic fluid. If HFB, HFC or HFD or environmentally acceptable hydraulic fluids are used, the limitations regarding technical data or other seals must be observed.

Details regarding the choice of hydraulic fluid

The correct choice of hydraulic fluid requires knowledge of the operating temperature in relation to the ambient temperature: in an open circuit, the reservoir temperature.

The hydraulic fluid should be chosen so that the operating viscosity in the operating temperature range is within the optimum range (v_{opt} see shaded area of the selection diagram). We recommended that the higher viscosity class be selected in each case.

Example: At an ambient temperature of X °C, an operating temperature of 60 °C is set in the circuit. In the optimum operating viscosity range (v_{opt}, shaded area), this corresponds to the viscosity classes VG 46 or VG 68; to be selected: VG 68.

Note

The case drain temperature, which is affected by pressure and speed, can be higher than the reservoir temperature. At no point of the component may the temperature be higher than 115 °C. The temperature difference specified below is to be taken into account when determining the viscosity in the bearing.

If the above conditions cannot be maintained due to extreme operating parameters, we recommend flushing the case at port U (sizes 250 to 1000).

Selection diagram

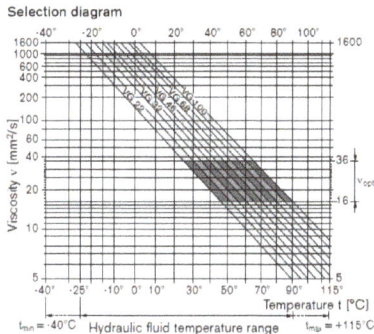

Fig. 4.30- Example of Manufacturer's Instructions about Hydraulic Fluid for A2FO Pump (Courtesy of Bosch Rexroth)

56

4.3.17- Review Prime Mover Overloading Conditions

❖ **General Prime Mover:**

- Ideally, it should drive the pump at the prim mover maximum efficiency.

- Practically, it should comfortably (with an acceptable efficiency) satisfy the power needed by the pump in all phases of operation.

- Review the power-efficiency curve of the prime mover.

❖ **Single-Phase Electric Motors:**

- They have less starting torque than 3-phase motors.

- It can be overloaded for short periods up 10% > rated power.

- Above that, circuit breaker trips the motor →

- Hydraulic circuit design should consider unloading the pump at start up.

57

❖ **Three-Phase Electric Motors:**

▪ They can be overloaded for short periods up 25% > rated power.

❖ **Engines:**

▪ Engine power is reduced over time due to engine aging.

▪ They can't work at a power above the rated power even momentarily →

▪ Hydraulic circuit design should consider pump unlading at start up to avoid stalling the engine.

58

58

4.3.18- Proper Placement of Hydraulic Power Unit

❖ Industrial applications → power unit is placed beside or overhead the machine.

❖ Mobile applications → pump is driven by an engine vis e power take-off shaft. Rest of the components are distributed within the machine.

❖ **Best Practices for Power Unit Installation:**

▪ **Visual Inspection:** Before installation, check for visible transport damage e.g. cracks, leaking seals, screws, protective covers.

▪ **Air Circulation:** A preferred location is where free air can circulate around all sides.

▪ **Use of Forced Air Ventilation if Needed:** Do not install a power unit inside a cabinet, console, under a table or bench, or other closed space where air cannot freely circulate unless forced air ventilation is provided.

59

59

- **Shield from Heat Sources:**

  o Power units shouldn't be exposed to extreme temperatures s, either hot or cold.

  o Do not install in direct sunlight without a sunshade, and one which will not restrict vertical air circulation.

  o Do not install near a furnace without interposing a heat shield unless a provision has been made to keep the oil cool with a heat exchanger.

60

4.3.19- Proper Installation of Hydrostatic Transmission (Courtesy of Womack)

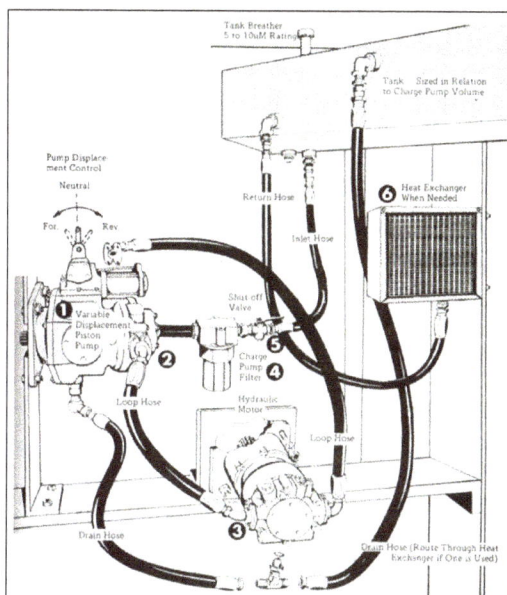

1. Main Pump.
2. Charge pump.
3. Hydraulic Motor.
4. Charge Pump Filter.
5. Shut-Off Valve.
6. Heat Exchanger.

Fig. 4.31- Proper Installation of Hydrostatic Transmission (Courtesy of Womack)

61

4.4- BP-Pumps-04-Standard Tests and Calibration

4.4.1- Testing of Fixed Displacement Pumps

- **ISO 4409:2007** specifies the test requirements.

- New pump is factory tested.

- Q_{th} = theoretical flow, Q_{new} versus working pressure p_p.

- A test characteristic curve (Q_{test}) must be developed.

- Internally drained pump $\rightarrow$ only choice is to measure pump outlet flow.

- Externally drained pump $\rightarrow$ measuring either pump outlet flow or the case drain (q_L).

- **Q-p** plot $\rightarrow$ volumetric efficiency.

- P_{in}-**p** $\rightarrow$ overall efficiency and mechanical efficiency.

62

62

$$\text{Volumetric Efficiency} = Q_{test} / Q_{th} = 1 - (q_L / Q_{th})$$
$$\text{Overall Efficiency} = P_{out} / P_{in} = (p_p \times Q_{test}) / P_{in}$$
$$\textbf{Mechanical Efficiency} = \textbf{Overall Efficiency} / \textbf{Volumetric Efficiency}$$

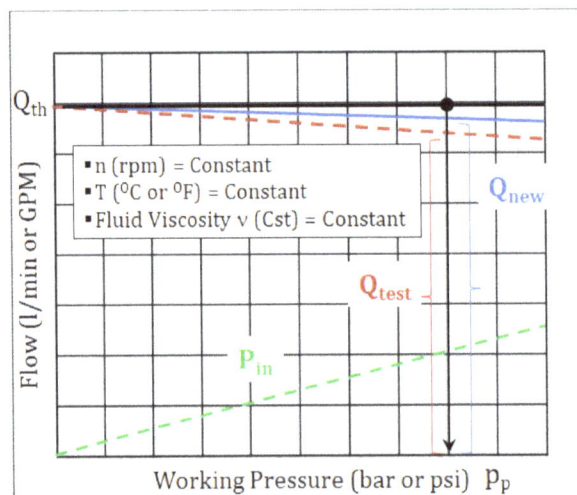

Fig. 4.32- Characteristic Curve for a Positive Displacement Pump

63

63

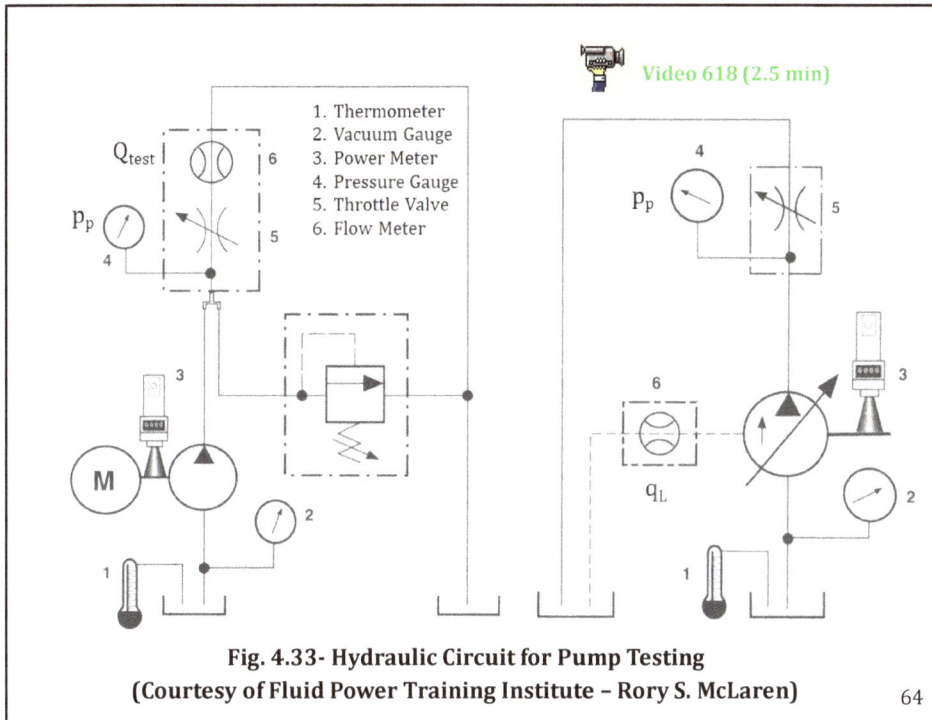

Video 618 (2.5 min)

1. Thermometer
2. Vacuum Gauge
3. Power Meter
4. Pressure Gauge
5. Throttle Valve
6. Flow Meter

Fig. 4.33- Hydraulic Circuit for Pump Testing
(Courtesy of Fluid Power Training Institute – Rory S. McLaren)

64

64

4.4.2- Setting of Pressure Compensated Pump

- Cracking (critical) pressure (p_{CR}) and cutoff pressure (p_{CO}).
- Backup pressure relief valve set 10 bar > pump compensator p_{CO}.

1. Turn OFF the power
2. Fully open the PRV
3. Fully close the pump compensator
4. Turn ON the power
5. Pressure now = tank pressure
6. Close the PRV gradually until $p_p = p_{CO} + 10$ bar
7. Open the pump compensator to lower pump pressure until $p_p = p_{CO}$.

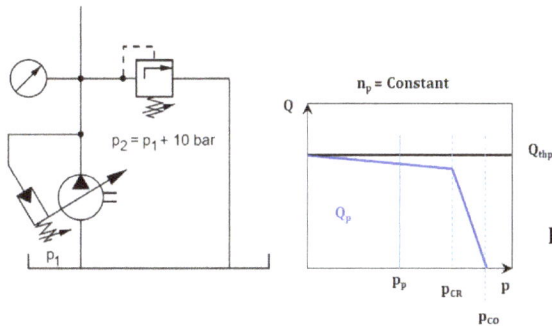

Fig. 4.34- Setting of a Pressure Compensated Pump

65

4.5- BP-Pumps-05-Transportation and Storage

4.5.1-Pump and Power Units Transportation

- A pump could be of a significant weight and size.
- Improper transportation → it loses stability fall.
- Review the instructions provided by the manufacturer.

Transport with ring screw

The drive shaft can be used to transport the axial piston unit as long as only outward (pulling) axial forces occur. Thus, you can suspend the axial piston unit from the drive shaft.

- To do this, screw a ring screw completely into the thread on the drive shaft. The threaded sizes is stated in the installation drawing.
- Make sure that each ring screw can bear the total weight of the axial piston unit plus approx. 20%.

You can hoist the axial piston unit as shown with the ring screw screwed into the drive shaft without any risk of damage.

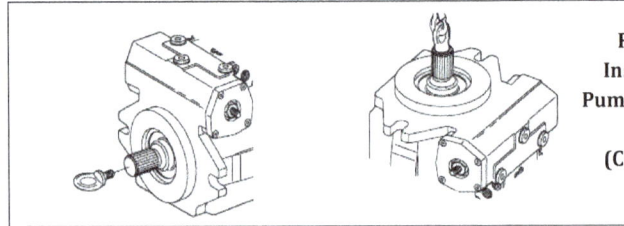

Fig. 4.35- Manufacturer's Instructions for Axial Piston Pump Transportation using Ring Screw (Courtesy of Bosch Rexroth)

66

Transport with lifting strap

- Place the lifting strap around the axial piston unit in such a way that it passes over neither the attachment parts (e.g. valves) nor such that the axial piston unit is hung from attachment parts

Risk of injury!

During transport with a lifting device, the axial piston unit can fall out of the lifting strap and cause injuries.

- Hold the axial piston unit with your hands to prevent it from falling out of the lifting strap.
- Use the widest possible lifting strap.

WARNING!

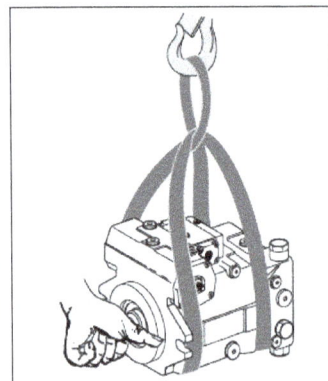

Fig. 4.36- Manufacturer's Instructions for Axial Piston Pump Transportation using Lifting Strap (Courtesy of Bosch Rexroth)

67

If no instructions are found:

- Ensure that no unauthorized persons are within the hazard zone.
- Check the weight and the center of gravity of the hydraulic power unit.
- Make certain that the lifting device has adequate lifting capacity.
- Only the attachment points should be used for lifting the product.
- Place the product on a suitable surface rated to the weight of the unit or on the ground.
- A hydraulic power unit must never be attached to or lifted at the mounted components (piping, hoses, manifolds, electric motors, accumulators, etc.).

68

68

4.5.2-Pump and Power Units Storage

- Review the storage instructions from the manufacturer

Storage Conditions	Packaging	Protective Agent	Storage Time (Months)	
			Test with the protective agent	Filling with the protective agent
Storage in dry rooms at constant temperature	For carriage overseas	Mineral oil	12	24
		Corrosion protection oil	12	24
	Not for carriage overseas	Mineral oil	9	24
		Corrosion protection oil	12	24
Outdoor storage (protect the product against damage and water ingress)	For carriage overseas	Mineral oil	6	12
		Corrosion protection oil	9	24
	Not for carriage overseas	Mineral oil	0	12
		Corrosion protection oil	6	24

Table 4.4- Manufacturer's Instructions for Storage of a Power Unit (Mcmillan Engineering Group)

69

69

Storing the axial piston unit

Requirement

- The storage areas must be free from corrosive materials and gasses.
- The storage areas must be dry.
- Ideal storage temperature: +5 °C to +20 °C
- Minimum storage temperature: -50 °C.
- Maximum storage temperature: +60 °C.
- Avoid intense lights.
- Do not stack axial piston units and store them shock-proof.
- Do not store the axial piston unit on sensitive attachment parts, e.g. sensors.
- For other storage conditions, see Table 4.

▶ Check the axial piston unit monthly to ensure proper storage.

Fig. 4.37- Manufacturer's Instructions for Storage of an Axial Piston Pump (Courtesy of Bosch Rexroth)

After delivery

The axial piston units are provided ex-works with corrosion protection packaging (corrosion protection film).

Listed in the following table are the maximum permissible storage times for an originally packed axial piston unit.

Table 4: Storage time with factory corrosion protection

Storage conditions	Standard corrosion protection	Long-term corrosion protection
Closed, dry room, uniform temperature between +5 °C and +20 °C. Undamaged and closed corrosion protection film.	Maximum 12 months	Maximum 24 months

70

70

If no instructions are found:

- **Pump Case:**
  o Pump case is to be filled with hydraulic oil or a rust preventive oil, which is compatible with the rubber seals.
  o Leave a small air space for the oil to expand if the pump should be exposed to heat.

- **Pump Ports:** Ports are to be plugged with suitable plugs or dust caps.

- **Pump Shaft:** The shaft and other external machined surfaces should be coated with grease.

- **Long term storage:** If the pump is going to be stored for more than a year, shaft seal should be replaced before putting it in service.

71

71

Best Practice of **Gear Pump** Maintenance and Inspection

Additional Links:

- https://www.youtube.com/watch?v=5quU4EbuDNs
- https://www.youtube.com/watch?v=A8_r1tgyLgE
- https://www.youtube.com/watch?v=lsuAC4w2520
- https://www.youtube.com/watch?v=VhEazvwukBQ

Video 440 (6 min)

Video 441 (5 min)

Media Type	Title	Minutes
	Video-128	External Gear Pump Basic Parts	1
	Video-131	Gerotor Basic Parts	0.5
	Video-38	External Gear Pump Maintenance	25
	Video-39	Internal Gear Pump Maintenance	11

72

72

Best Practice of **Vane Pump** Maintenance and Inspection

Media Type	Description	Minutes
	Video-152	Vane Pump Maintenance	8
	Video-173	Cartridge Replacement	1

73

73

Best Practice of **Piston Pump** Maintenance and Inspection

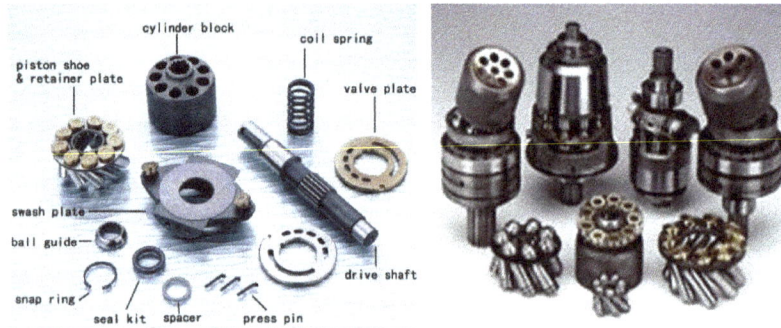

Media Type		Description	Minutes
	Video-129	Axial Piston Swash Plate Pump	1
	Video-130	Axial Piston Bent Axis Pump	1

74

74

Chapter 4 Reviews

1. The installed pump shown below will be subject to?
   A. Aeration and cavitation.
   B. Noise and vibration.
   C. Reduced pump flow.
   D. All the above.

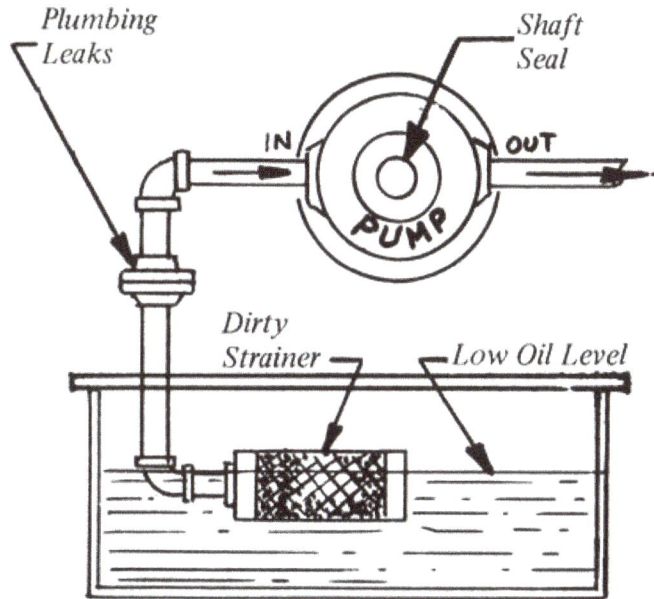

2. The pump shown below is conventionally considered?
   A. Unidirectional pump running "Left".
   B. Unidirectional pump running "Right".
   C. Bidirectional pump.
   D. No indication of the shaft rotation.

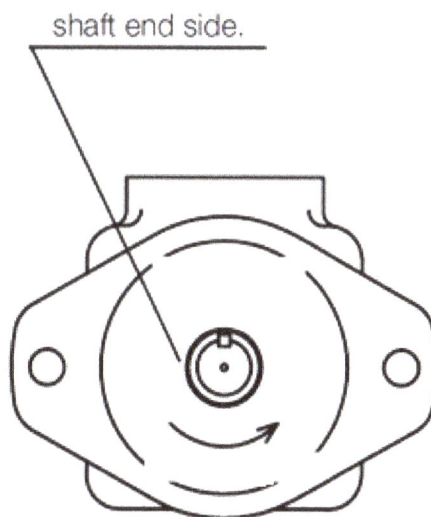

3. The direction of rotation of the balanced vane pump shown below is?
   A. Counter-clockwise.
   B. Clockwise.
   C. Bidirectional.
   D. This Pump assembly has nothing to do with the direction of rotation

Note Position of Cam Lobes

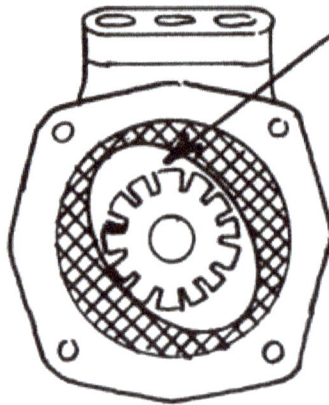

4. The direction of rotation of the unbalanced vane pump shown below is?
   A. Left.
   B. Right.
   C. Bidirectional.
   D. This Pump assembly has nothing to do with the direction of rotation

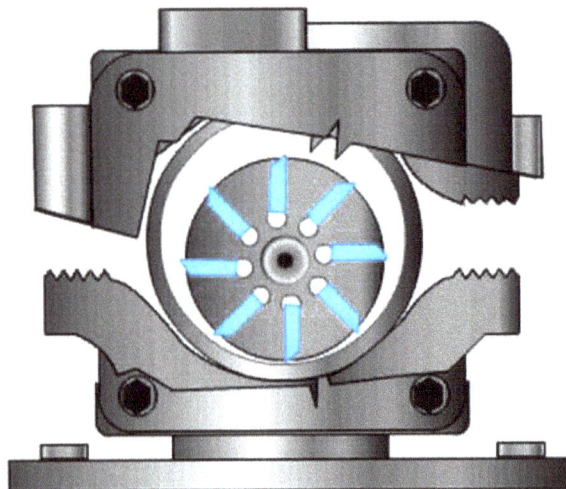

5. The drive gear of the gear pump shown below rotates?
   A. Counterclockwise.
   B. Clockwise.
   C. Bidirectional.
   D. This Pump assembly has nothing to do with
      the direction of rotation

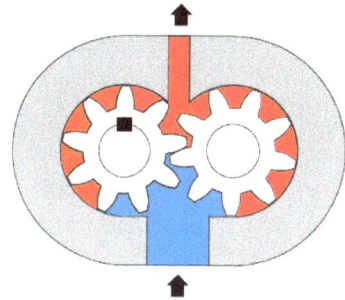

6. IF an unbalanced pump is driven by a belt as shown below, the pump should be oriented so
   that the outlet port faces the motor in order to?
   A. Get more torque from the motor.
   B. Increase the pump pressure.
   C. Balance the load on the bearing from both sides.
   D. Leverage the pump speed.

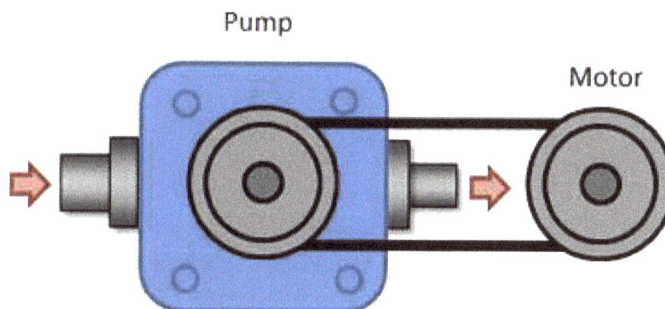

7. Alignment of pump and drive motor shafts is an important requirement. By placing the dial
   indicator as shown below, which of the following is measured?
   A. Radial misalignment between pump and motor shafts.
   B. Axial misalignment between pump and motor shafts.
   C. Face clearance of the coupling.
   D. None of the above.

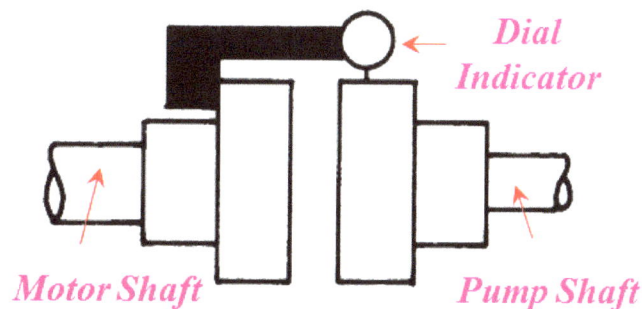

8. In the system shown below, a pressure relief valve has been used as a backup in case the pump compensator fails to work. Which of the following statements is correct?
   A. PRV set lower than the pump compensator. PRV is to be adjusted before then the pump compensator.
   B. PRV set higher than the pump compensator. PRV is to be adjusted before the pump compensator.
   C. PRV set lower than the pump compensator. PRV is to be adjusted after the pump compensator.
   D. PRV set higher than the pump compensator. PRV is to be adjusted after the pump compensator.

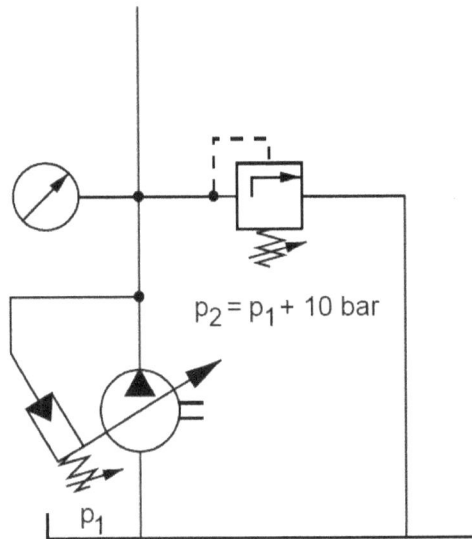

$$p_2 = p_1 + 10 \text{ bar}$$

p_1

9. When a pump and drive motor are assembled using bell housing, which of the following statement is considered FALSE?
   A. Alignment should be repeated every time the pump is removed.
   B. Alignment does not need to be repeated every time the pump is removed.
   C. Distance between shafts is considered in the design of the bellhousing.
   D. All the above statements are false.

10. An acceptable case drain pressure is approximately.
    A. 10 psi.
    B. 25 psi.
    C. 75 psi.
    D. 125 psi.

Chapter 4 Assignment

Student Name: -- Student ID: -------------------

Date: -- Score: ------------------------

Question: Explain the process of setting a compensator of a pressure compensated pump when an additional pressure relief valve is used as a backup.

Chapter 5
Maintenance of Motors

Objectives:

This chapter provides guidelines for **motors** selection, replacement, maintenance scheduling, installation, testing, storage and transportation. This chapter is supported by examples and figures granted by leading fluid power manufacturers.

Brief Contents:

5.1-BP-Motors-01-Selection and Replacement
5.2-BP-Motors-02-Maintenance Scheduling
5.3-BP-Motors-03-Installation and Maintenance
5.4-BP-Motors-04-Standard Tests and Calibration
5.5-BP-Motors-05-Transportation and Storage

0

0

5.1- BP-Motors-01-Selection and Replacement
5.1.1- Selecting or Replacing Motors

When selecting or replacing an existed motor →
BP-Motors-01-Selection and Replacement:
- Review maximum/optimum operating pressure.
- Review min/maximum/optimum operating speed.
- Review maximum operating torque.
- Review maximum overall efficiency at the optimum operating conditions.
- Review size (**See Note 1**) and displacement control requirements.
- Review type of fluid.
- Review contamination tolerance.
- Review noise level.
- Review initial cost.
- Review approximate service life.
- Review availability and interchangeability.
- Review maintenance and spare parts.
- Review physical size and weight.

1

1

Note 1: if a motor is <u>undersized</u> →

- Motor speed increases → motor work inefficiently.

- To maintain same external torque → working pressure increases →

- Maximum system pressure reset.

5.1.2- Displacement Calculation of Legacy Motors

Equations 4.1 through 4.8 (in Chapter 4 for pumps) are applicable for finding displacements of various motors mechanisms.

2

5.2- BP-Motors-02-Maintenance Scheduling

#	Preventive Maintenance Actions	Daily	Weekly	Monthly	Biannually	Annually
1	Clean around and outside surface	✓	✓	✓	✓	✓
2	Check for unusual sound	✓	✓	✓	✓	✓
3	Check temperature of the motor body	✓	✓	✓	✓	✓
4	Check tightness and leakage around hydraulic connections		✓	✓	✓	✓
5	Check electrical connections (if found)		✓	✓	✓	✓
7	Check for vibration and condition of dampers			✓	✓	✓
8	Standard tests and calibration					✓

Table 5.1- BP-Motors-02-Maintenance Scheduling

3

5.3- BP-Motors-03-Installation and Maintenance

- Pumps and motors have the same mechanisms →
- Installation and maintenance guidelines are same.
- Motor cavitation → consider using anti-cavitation check valves.

BP-Motors-03-Installation and Maintenance:
1. Install the motor to avoid Cavitation.
2. Identify Ports and Direction of Rotation.
3. Proper Shaft Alignment for Direct Drive.
4. Adequately Dampen Vibration.
5. Proper Priming.
6. Proper Case Drain.
7. Review Range of Rotational Speed.
8. Review Range of Working Pressure.
9. Review Range of Working Temperature.
10. Review Compatibility with Working Hydraulic Fluid.

4

4

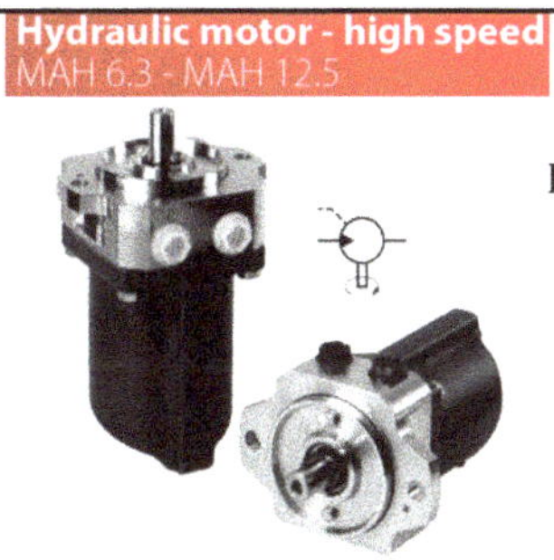

Hydraulic motor - high speed
MAH 6.3 - MAH 12.5

**Fig. 5.1- Manufacturer's Instructions for
Hydraulic Motor Installation
(Courtesy of Danfoss)**

Drain line

Max. pressure = 6 bar absolute.
Drain pressure must never exceed return pressure
by more than 1 bar.

Installing the drain line
The drain line/motor must be positioned so that
the motor cannot empty itself during standstill.

Temperature

Fluid temperature:
Min. +3°C to max. +50°C. at max. pressure
Min. +3°C to max. +60°C. at max. 100 bar

In case of lower operating temperatures,
please contact the Danfoss Sales
Organization for Water Hydraulics.

Ambient temperature:
Min. 0°C to max. 50°C.

Storage temperature:
Min. -40°C to max. +70°C.

Motor variants

MAH motors are optimized for operation in one
direction and are therefore available in CW and
CCw versions.

Filtration

The water supplied to the valve must be filtered:
10 μm absolute, ß$_{10}$-value > 5000 filter is
recommended.

For further information on filters, please
contact the Danfoss sales department for water
hydraulics.

5

5

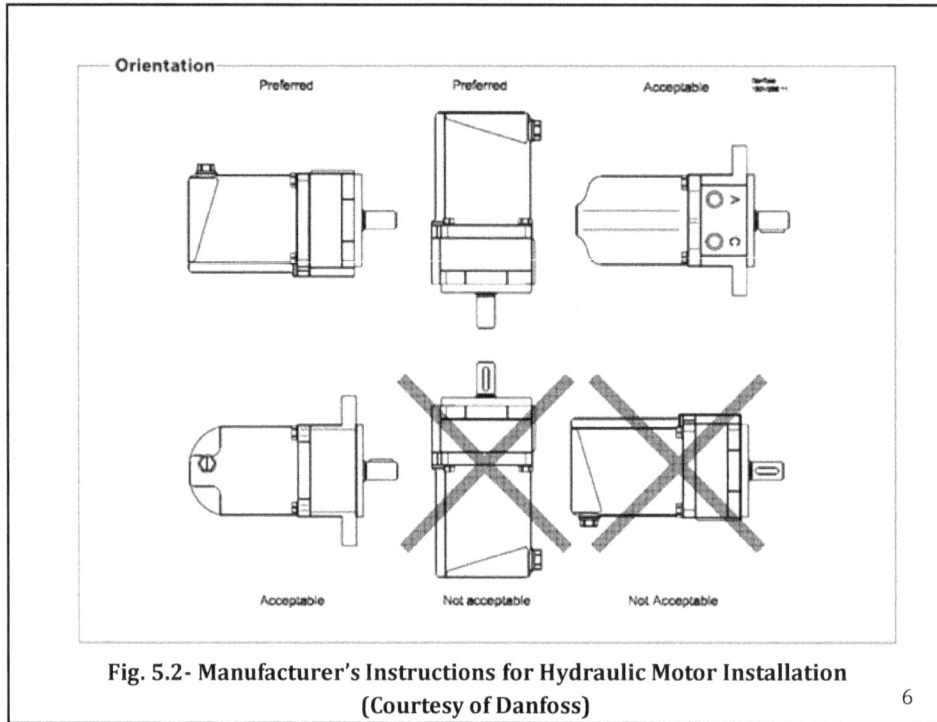

Fig. 5.2- Manufacturer's Instructions for Hydraulic Motor Installation
(Courtesy of Danfoss)

6

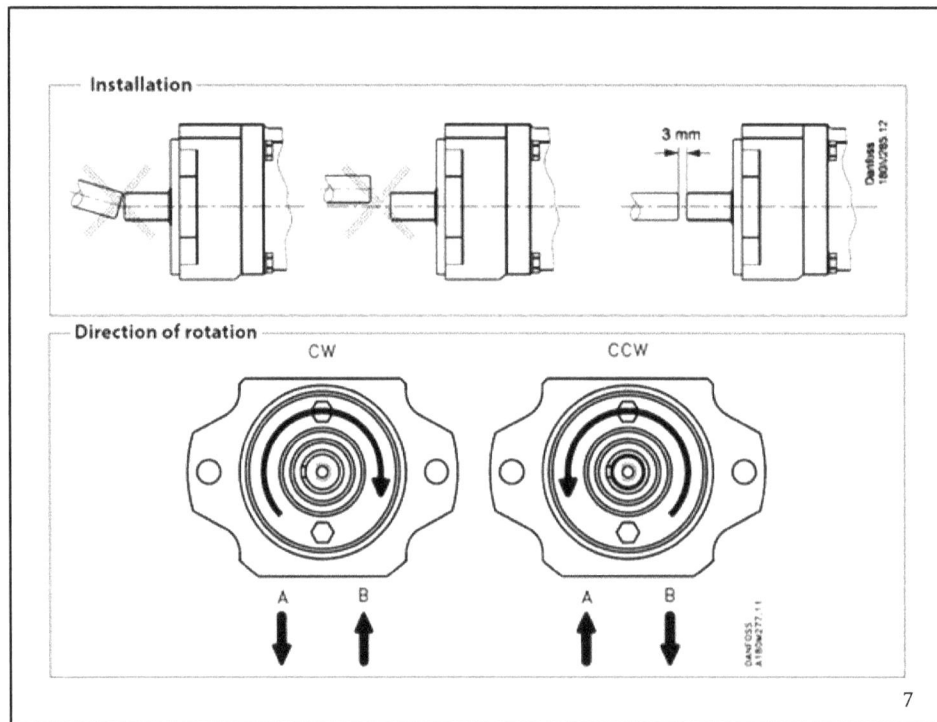

5.4- BP-Motors-04-Standard Tests and Calibration

- **SAE-J746** specifies the test requirements.

- New motor is factory tested.

- n_{th} = theoretical speed = f (motor size and inlet flow).

- n_{test} that is slower because of internal leakage.

- T_{th} = theoretical torque = f (motor size & differential pressure).

- T_{test} is lower because of internal friction.

Volumetric Efficiency = n_{test} / n_{th}
Mechanical Efficiency = T_{test} / T_{th}
Overall Efficiency = Volumetric Efficiency x Mechanical Efficiency

For more detailed discussions about motor sizing and efficiency calculations, refer to Volume 1 of this series of textbooks "Introduction to Hydraulics for Industry Professionals".

8

8

Video 670 (1.5 min)

Video 671 (2 min)

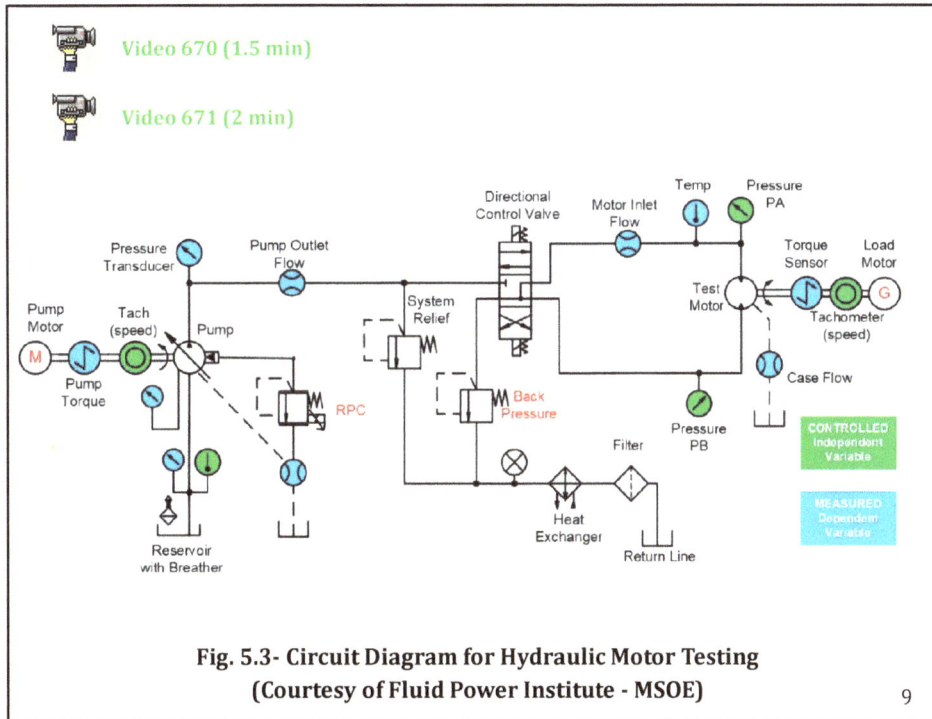

**Fig. 5.3- Circuit Diagram for Hydraulic Motor Testing
(Courtesy of Fluid Power Institute - MSOE)**

9

9

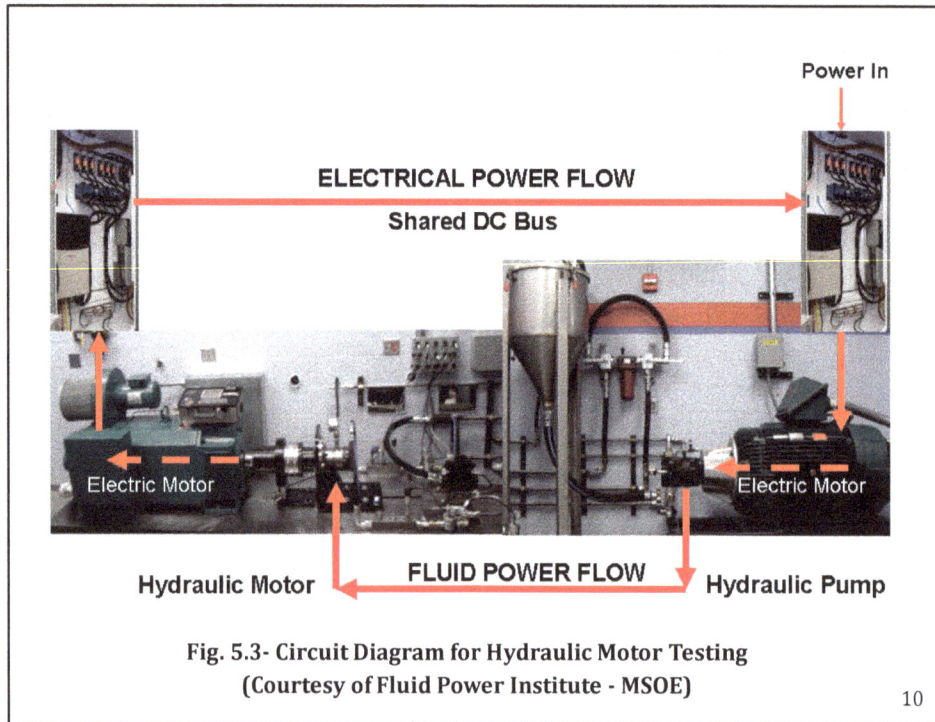

**Fig. 5.3- Circuit Diagram for Hydraulic Motor Testing
(Courtesy of Fluid Power Institute - MSOE)**

10

10

5.5- BP-Motors-05-Transportation and Storage

Same guidelines for storage and transportation of hydraulic pumps are applicable for hydraulic motors. So, refer to section 4.5.2. in Chapter 4 of this book

11

11

Chapter 5 Reviews

1. If a larger motor is used to replace a defective smaller motor, assuming the supply flow and loading conditions are the same, what would be the expected changes in the system performance?
   A. Increased motor RPM and increased working pressure.
   B. Increased motor RPM and decreased working pressure.
   C. Decreased motor RPM and increased working pressure.
   D. Decreased motor RPM and decreased working pressure.

2. Which of the following ratios represents a hydraulic motor volumetric efficiency?
   A. (Output Power) / (Input Power).
   B. (Actual RPM) / (Theoretical RPM)
   C. (Actual Torque) / (Theoretical Torque)
   D. (Actual Temperature) / (Theoretical Temperature)

3. Which of the following ratios represents a hydraulic motor mechanical efficiency?
   A. (Output Power) / (Input Power).
   B. (Actual RPM) / (Theoretical RPM)
   C. (Actual Torque) / (Theoretical Torque)
   D. (Actual Temperature) / (Theoretical Temperature)

4. Which of the following ratios represents a hydraulic motor overall efficiency?
   A. (Output Power) / (Input Power).
   B. (Actual RPM) / (Theoretical RPM)
   C. (Actual Torque) / (Theoretical Torque)
   D. (Actual Temperature) / (Theoretical Temperature)

5. Among the preventive maintenance actions shown below, the most frequent action is?
   A. Perform motor standard tests.
   B. Checking vibration and condition of dampers.
   C. Clean around and the outside surfaces of the motor.
   D. Check electrical connection (if found).

Chapter 5 Assignment

Student Name: --- Student ID: ------------------

Date: --- Score: ------------------------

Question: list the design and operating design parameters that should be reviewed when replacing a hydraulic motor.

Chapter 6
Maintenance of Cylinders

Objectives:

This chapter provides guidelines for **cylinders** selection, replacement, maintenance scheduling, installation, testing, storage and transportation. This chapter is supported by examples and figures granted by leading fluid power manufacturers.

Brief Contents:

6.1-BP-Cylinders-01-Selection and Replacement

6.2-BP-Cylinders-02-Maintenance Scheduling

6.3-BP-Cylinders-03-Installation and Maintenance

6.4-BP-Cylinders-04-Standard Tests and Calibration

6.5-BP-Cylinders-05-Transportation and Storage

0

0

6.1- BP-Cylinders-01-Selection and Replacement

When selecting or replacing an existed cylinder →

- **Cylinder Size:**
  o Cylinder size changed → speed and pressure of the new cylinder must be checked to determine if it will provide acceptable machine operation.

- **Cylinder Cushioning:**
  o Newer cylinder must be equipped with same cushioning as the old one.
  o Cushioning throttle valve should be reset at the time of installation.

- **Cylinder Installation Requirements:**
  o Review mounting the cylinder with the machine body.
  o Review engagement with the load.
  o Review connection to transmission lines.

1

1

6.2- BP-Cylinders-02-Maintenance Scheduling

Unless otherwise stated by components and systems manufacturer:

#	Preventive Maintenance Actions	Daily	Weekly	Monthly	Biannually	Annually
1	Clean the outer surface of the cylinder barrel + Clean around and the ports.	✔	✔	✔	✔	✔
2	Check if there is any sign of damage or even rubbing in the barrel paint.	✔	✔	✔	✔	✔
3	Check if there is any sign of leaking.		✔	✔	✔	✔
4	Check proper connection with the load.		✔	✔	✔	✔
5	Check proper connection to plumbing.			✔		
6	Check Surface temperature. Air bleeding.			✔	✔	✔
7	Greasing the cylinders connection points with the external loads.			✔	✔	✔
8	Cylinders maintenance and routine inspection from inside.				✔	✔

Table 6.1- BP-Cylinder-02-Maintenance Scheduling

2

6.3- BP-Cylinders-03-Installation and Maintenance

- **ISO/TS 13725** provides detailed information about cylinder installations.

BP-Cylinders-03-Installation and Maintenance:
1. Proper Cylinder Disassembling.
2. Proper Cylinder Inspection and Maintenance.
3. Proper Seal Replacement and Installation.
4. Proper Cylinder Assembling
5. Proper Mounting on the Machine Structure.
6. Proper Alignment with the Load.
7. Proper Air Bleeding.
8. Proper Connection with Transmission Lines.
9. Propper Installation of External Limit Switches.
10. Protect End Caps Against Impact Load.
11. Protect Cylinder from External Hazard.
12. Protect Cylinder Rod from Corrosion.
13. Protect Air Chamber of a Single-Acting Cylinder from the Environment.
14. Review Range of Allowable and Maximum Working Conditions.

3

6.3.1- Proper Cylinder Disassembly

- Practices of removing the cylinder from the machine is out of scope of this textbook.
- After removing the cylinder from the machine →

❖ **Know the Component:**

- Major parts of a hydraulic cylinder.

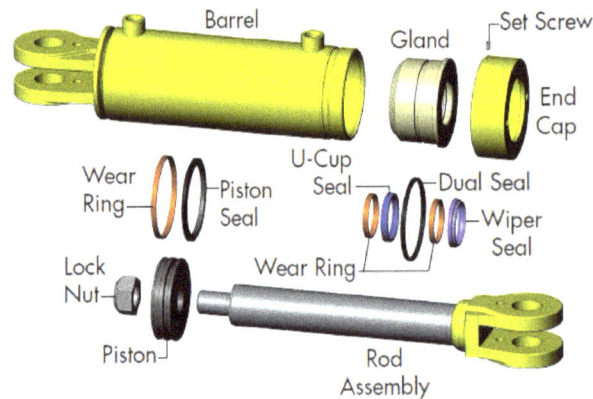

Fig. 6.1- Typical
Construction of a
Double Acting Cylinder
(www. degelman.com)

4

4

❖ **Follow Proper Disassembly Procedure:**

- Example of guidelines for disassembling mill-type cylinder that has threaded end caps:

1. Loosen Set Screw and turn off end cap.
2. Carefully remove piston/rod/gland assemblies and place them on an appropriate fixture that won't damage the rod surface finish and plating.
3. Disassemble the piston from the rod assembly by removing lock nut.
4. DO NOT clamp rod by chrome surface.
5. Slide off gland assembly & end cap.
6. Remove seals and inspect all parts for damage.
7. Install new seals and replace damaged parts with new components.
8. Inspect the inside of the cylinder barrel, piston, rod and other polished parts for burrs and scratches.

5

❖ **Avoid Oil Spillage:**

▪ Use a drip pan or other mean to avoid oil spillage during disassembly.

▪ Allow sufficient time to drain all fluid before pulling the cylinder rod out of the cylinder.

❖ **DO NOT Underestimate the Power:**   ❖ **Proper Component Support:**

V273 (2.5 min)

Fig. 6.2- Improper Cylinder Disassembling
(Courtesy Fluid Power Safety Institute)

Fig. 6.3- V-Blocks to Support Cylinder Rods
(www.ame.com)

6

6

6.3.2- Proper Cylinder Inspection and Maintenance

❖ **Cylinder Barrel Inspection:**

▪ Visual inspection from inside using flashlight.

▪ Scratches →
  ○ Cylinder operate sluggishly.
  ○ Cylinder lose power.
  ○ Load drifts.

▪ Small scratches → barrel can be honed.

▪ Harsh scratches → barrel is replaced.

▪ Critical applications → checked for barrel straightness and out of round.

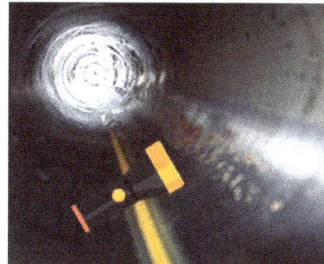

Fig. 6.4- Inspecting a Cylinder Barrel

7

7

❖ **Cylinder Rod Inspection:**

- A bent rod → excessive rod bearing wear and seal deterioration.
- A damaged seal → leakage + loss of power + premature failure.
- Manufacturer → allowable run-out.
- However, 0.5 mm/meter is considered an acceptable value.
- Rod straightness must routinely be inspected.
- Small run out → rod can be straightened on a press.
- Rod exceeds allowable run-out → rod is replaced.

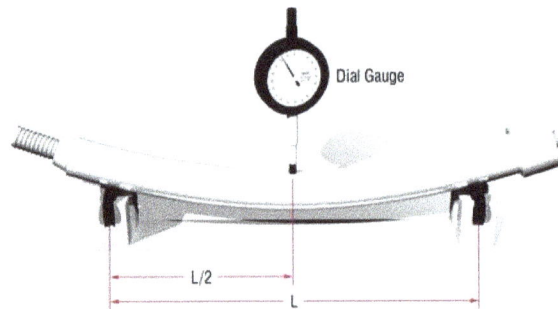

Fig. 6.5 – Checking Rod Straightness (www.machinerylubrication.com) 8

8

6.3.3- Proper Seal Replacement and Installation

❖ **In case of seal replacement, review:**

- Seal Type.
- Seal Dimensions.
- Seal Lip Geometry.
- Seal Cross Section.
- Seal Material Based on Working Temperature.
- Seal Material Based on Working Pressure.
- Seal Material Based on Working Fluid.
- Seal Material Based on Hardness.
- Seal Material Based on General Properties.

9

9

❖ **Best Practices for Hydraulic Seals Installation:**

1. **Properly Remove the Old Seal:** Carefully dismount the old seal without damaging the bores or shafts.

Fig. 6.6 - Properly Remove the Old Seal

2. **Use Genuine Seals:** DO NOT use non-branded seals or seals that are not approved by the hydraulic component manufacturer.

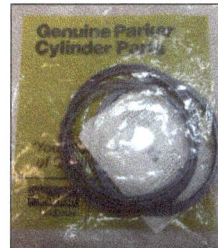

Fig. 6.7 - Use Genuine Seals (Courtesy of Parker)

10

10

3. **Never use a Pretensioned or Pre-Used Seal:**
- Used seal → plastic deformation + defected geometrical shape changes
- Even, are not seen by naked eyes.

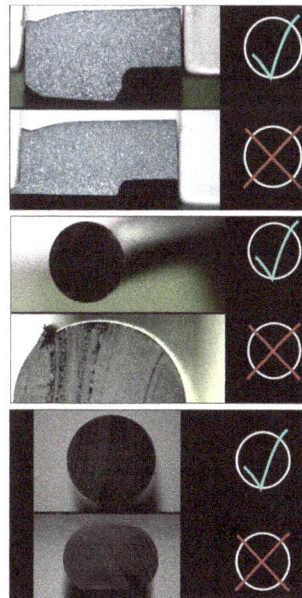

Fig. 6.8 - Never use a Pretensioned or Pre-Used Seal (Courtesy of Trelleborg)

11

11

4. Inspect New Seals:

- Inspect for damage on (circumference + sealing lip + outer diameter).
- DO NOT shorten the original tension spring if found.
- Double check the correct placement direction of the seal.

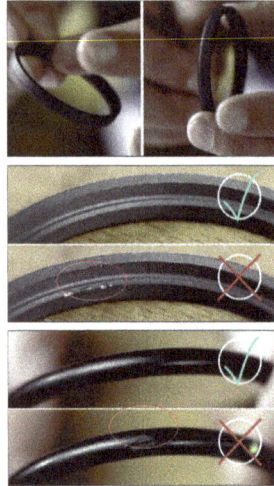

Fig. 6.9 - Inspect New Seals (Courtesy of Trelleborg)

12

12

5. Inspect Seal Groove: It should be clean and free of damage or sharp edges.

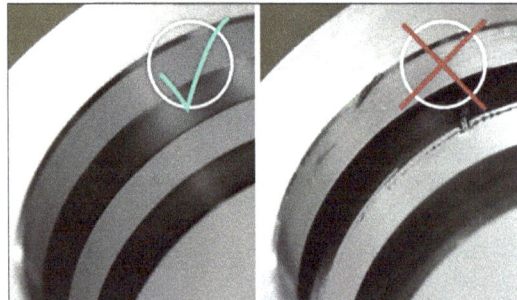

Visual inspection of the housing

Fig. 6.10 - Inspect Seal Groove (Courtesy of Trelleborg)

13

13

6. Inspect Assembly Tools: make sure they are clean and free of sharp edges, scratches, and contamination.

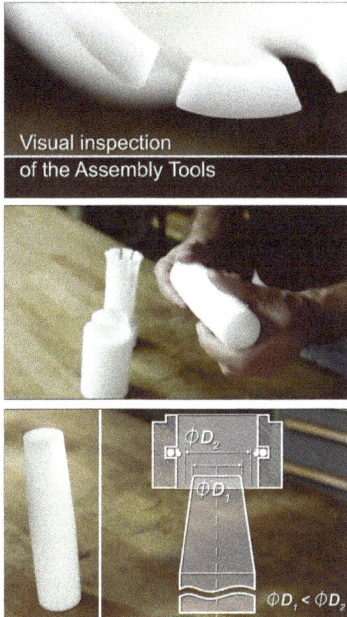

**Fig. 6.11 - Inspect Assembly Tools
(Courtesy of Trelleborg)**

14

14

7. Cover Threads: If the seal must be stretched over sharp edges, such as threaded parts or lead-in chamfers, these areas must be covered to prevent damaging the seals during the assembly process.

Fig. 6.12 - Cover Threads (Courtesy of Trelleborg)

15

15

8. **Lubricate Seal Before Installation:** Lubricate the seal and the mounting surface before seal installation. That reduces the surface friction on the seal and makes it easy to install it. Do NOT use regular grease. Use the same hydraulic fluid or a predefined compatible seal lubricant. Make sue it is clean.

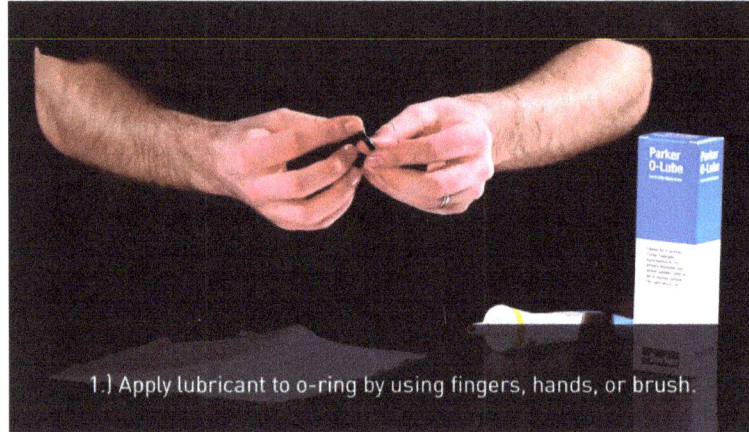

1.) Apply lubricant to o-ring by using fingers, hands, or brush.

Fig. 6.13 - Lubricate Seal Before Assembly (Courtesy of Parker) 16

16

9. **Use Adequate Installation Tools:** Use adequate tools for installing rod and piston seals to prevent damaging or twisting the seals.

Fig. 6.14.A - Use Adequate Installation Tools for Rod Seals (Courtesy of Trelleborg)

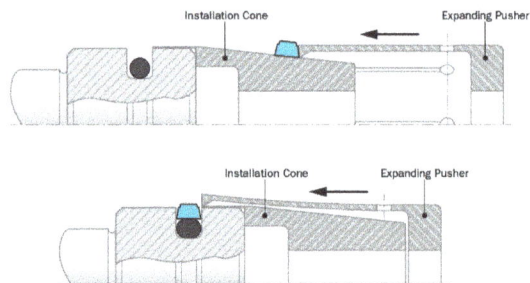

Fig. 6.14.B - Use Adequate Installation Tools for Piston Seals (Courtesy of Trelleborg) 17

17

10. Compress the Seal in kidney Shape: If needed or no assembly tools are found, compress the seal into a kidney shape. If the seal has a notch, DO NOT bend it from the position of the notch as this may cause overstretch or damage to the seal material.

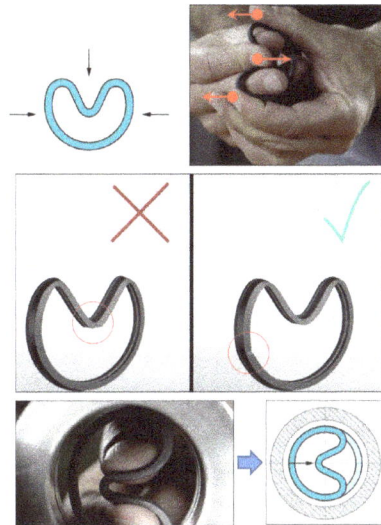

Fig. 6.15 - Compress the Seal in kidney Shape (Courtesy of Trelleborg)

18

18

11. Check Proper Installation of the Seal: make sure the seal is not tilted or twisted, and the seal axis coincides with the shaft or piston axis.

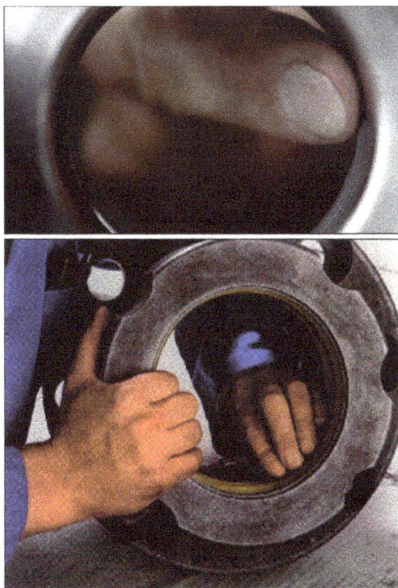

Fig. 6.16 - Check Proper Installation of the Seal

19

19

12. **Squeeze the Seal in its Groove:** In both static and dynamic applications, a certain amount of squeeze or compression is required upon installation to maintain contact with the sealing surfaces and prevent fluid leakage. This can be done by installation cones.

Fig. 6.17 - Squeeze the Seal in its Groove (Courtesy of Trelleborg)

V324-Rod Seal Installation (2 min)

V253-Cylinder Seals Installation (1.5 min)

20

20

6.3.4- Proper Cylinder Assembly

The following bullets provide guidelines for *assembling* the *cylinder*:

- **Cleaning:** Disassembled parts of the cylinder should be thoroughly cleaned using a solvent that is clean to standard and compatible with seals and parts.

- **Drying:** Cleaned parts should be dried by blowing with clean dry compressed air.

- **Lubricate:** Coat all parts with clean hydraulic fluid during assembly. Lubricating fluid should be same as the fluid that is used to fill the cylinder. No greases should be used.

21

6.3.5- Proper Mounting with the Machine Structure

- **Lifting and Transportation:** Rules are described in section 6.5.
- **Foot-Mounted Cylinders**
  - Supporting Surface: (straight, flat, leveled, and not twisting).
  - Pinning High Pressure Side: (carry shear force).
  - Support Long Cylinder Barrel: (avoid bending of the cylinder barrel)

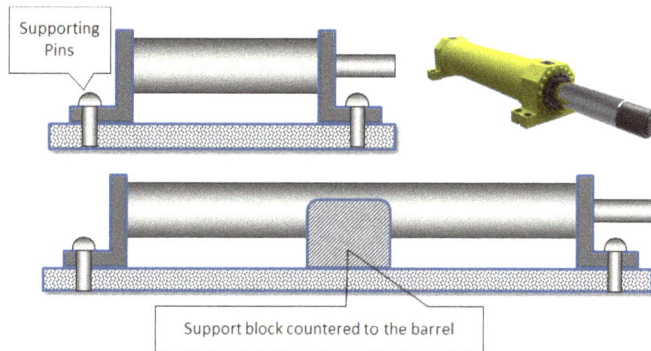

Fig. 6.18 – Foot-Mounted Cylinders   22

22

- **Flange-Mounted Cylinders:**
  - Rear Flange Mount: (buckling + reduced maximum allowable load).
  - Front Flange Mount: (recommended for high compressive load).

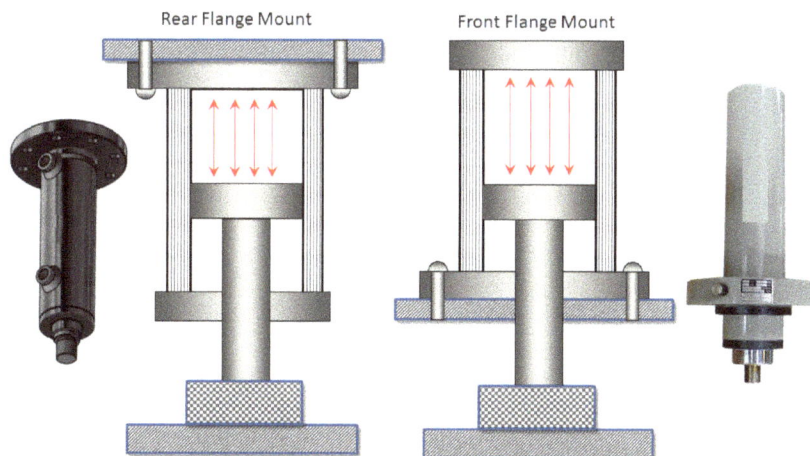

Fig. 6.19 – Flange-Mounted Cylinders   23

23

- **Eye-Mounted Clevis [Hinge-Mounted] Cylinders:**
  - o Cylinder Motion: Allow cylinder kinematic movement.
  - o Cylinder Connections: Flexible hoses or swivel connections.

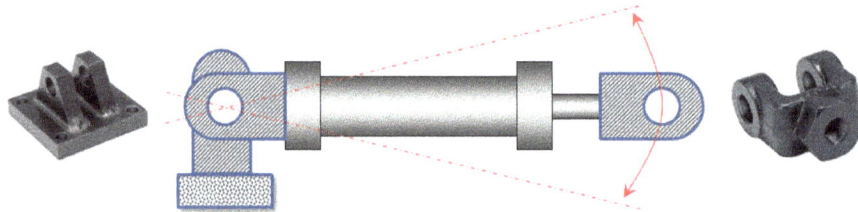

Fig. 6.20 – Eye-Mounted Cylinders

24

24

- **Trunnion-Mounted [Hinge-Mounted] Cylinders:**
  - o Cylinder Motion: Allow cylinder kinematic movement.
  - o Cylinder Connections: Flexible hoses or swivel connections.
  - o Trunnion Bearing: should be as close as possible to the cylinder barrel to reduce stresses on the trunnions.

Fig. 6.21 – Trunnion-Mounted Cylinders

25

25

6.3.6- Proper Alignment with the Load

- Misalignment → lateral load, seals failure, external leakage, cylinder rod damage, and cylinder buckling.
- Best practices for cylinder rod alignment with the load:
  - Decouple the load from the cylinder rod.
  - Make sure that the eye connections attached easily with without binding or transition fit.
  - Make sure that the hinge pin is inserted without binding, forcing or hammering.
  - Use shims, if needed, under the front and/or rear foot to help align the cylinder rod with the load.

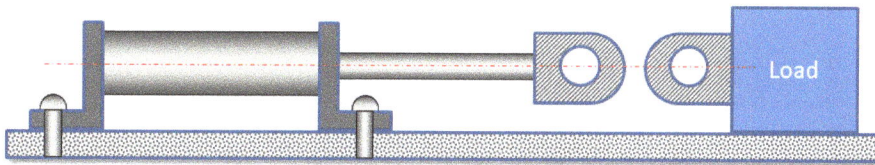

Fig. 6.22 – Cylinder Rod Alignment with the Load

26

26

6.3.7- Proper Air Bleeding

Why Air Bleeding:

- Air is compressed and suddenly expands →
- Piton and rod moves unexpectedly very fast→
- Possible end caps damages and risk of bodily injuries.
- Air compression → actuators move erratically.
- Air compression → dieseling effect damages the seals.

Best Practices for Hydraulic Cylinders air Bleeding:

- Make it part of regular maintenance.
- Perform it before installing a new hydraulic cylinder.
- Review instructions provided by the cylinder manufacturer.
- Make sure the hydraulic cylinder is not pressurized in any circumstances.
- Make sure to lock load to avoid free falling and risk of damage.
- Cylinder is positioned horizontally → bleeder valve is pointing upwards.
- Do not confuse air bleed ports with cushioning adjustment screws.
- Careful opening of the cylinder ports → avoid hazardous situation.

27

27

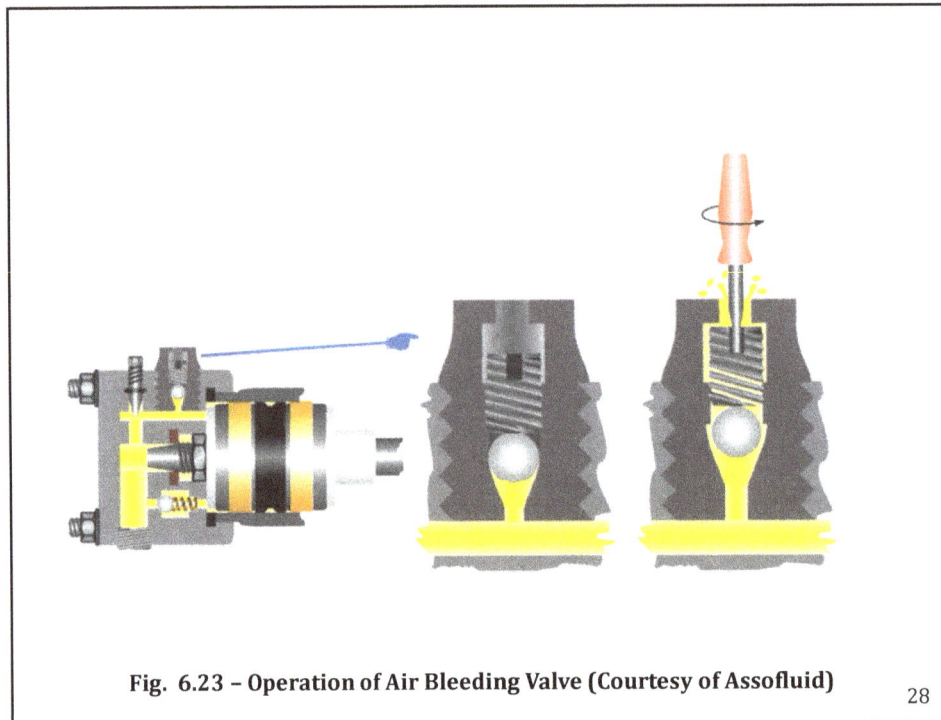

Fig. 6.23 – Operation of Air Bleeding Valve (Courtesy of Assofluid)

28

28

Automatic air bleeding Valve:

- Are available in the marketplace.
- Must be checked periodically following the manufacturer instructions.
- The valve consists of two parts.
- Lower part is permanently assembled in the cylinder.
- In bleeding and filling, only the upper part of the valve is opened.

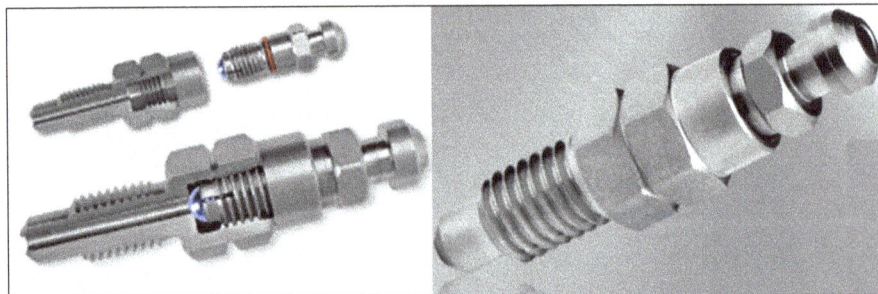

Fig. 6.24 – Automatic Air Bleeding Valve (www.stahlbus.com)

29

29

Examples of instructions provided by a manufacturer for air bleeding:

Bleeding Air from Double-Acting Type Hydraulic Cylinders:
1. Check all hoses or pipes are connected properly.
2. Fully retract the cylinder by working fluid.
3. Open the air valve at the piston side of the hydraulic cylinder.
4. Set up the hydraulic system and start it up.
5. Extend the piston rod slowly with no pressure built up.
6. Keep extending the piston rod until there is only oil (no foam) come out of the valve.
7. Shut down the system and close the air valve.
8. Depressurize the system.
9. Opening air bleeding valve at the rod side of the hydraulic cylinder.
10. Startup the hydraulic system.
11. Retract the hydraulic cylinder slowly with no pressure built up.
12. Keep retracting the cylinder until there is only oil (no foam) coming out of the valve.
13. Shut down the system and close the air valve.
14. Retest the system and cycle the hydraulic cylinder until it is running smoothly.

30

30

Bleeding Air from Single-Acting or Lift Dump Truck Cylinders:
1. Raise the dump body and cylinder to full extension. Leave the dump body in this position for several minutes to allow air to rise to the top of the cylinder.
2. Lower the dump truck cylinder until the front of the dump body is approximately 2 feet off the chassis frame.
3. Hold the dump body in this position.
4. Crack the air bleeding valve open.
5. Wait until all trapped air has escaped from the valve and a full stream of hydraulic oil is escaping from the valve.
6. At this point, the cylinder is bled, and the bleeder valve can be closed.

Fig. 6.25 – Single-Acting Cylinder in Dump Trucks (www.hydrauliccylindersinc.com)

31

31

6.3.8- Proper Connection with Transmission Lines

- Direct connection to the cylinder ports.
- Connection to a block mounted on the cylinder.
- Connection to a valve that is directly mounted on the cylinder barrel.

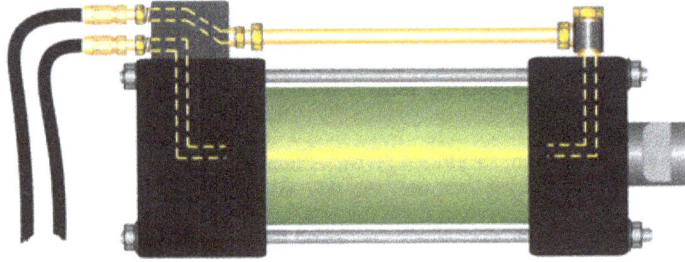

**Fig. 6.25 – Hydraulic Lines Connection
with the Cylinder (Courtesy of Assofluid)**

32

32

6.3.9- Proper Installation of External Limit Switches

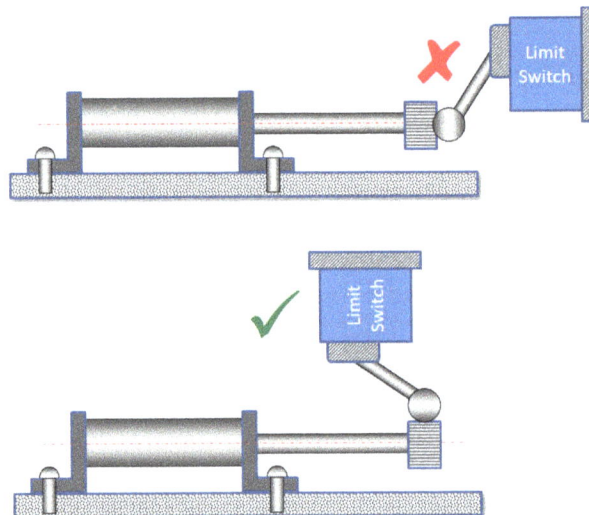

**Fig. 6.27 – Best Practices for Electro-Mechanical Limit
Switches Installation Against Cylinder Rod**

33

33

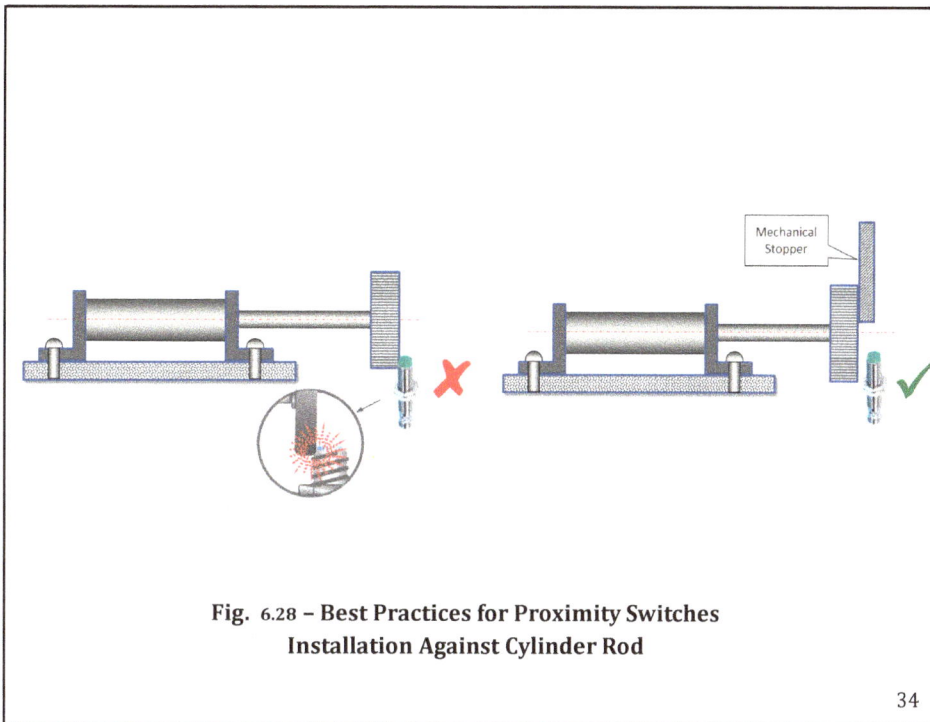

Fig. 6.28 – Best Practices for Proximity Switches Installation Against Cylinder Rod

34

6.3.10- Protect End Caps Against Impact Load

- If piston hits the end caps → fatigue failure.
- Cylinder speed < 10 m/s (4 in/s) → no cushioning actions is required.
- Cylinder speed = 10-20 cm/s (4-8 in/s → built in cushioning is required.
- Cylinder speed > 20 cm/s (8 in/s) → built in cushioning is not sufficient.
- External hydraulic deceleration system.
- External braking system or a shock absorber.

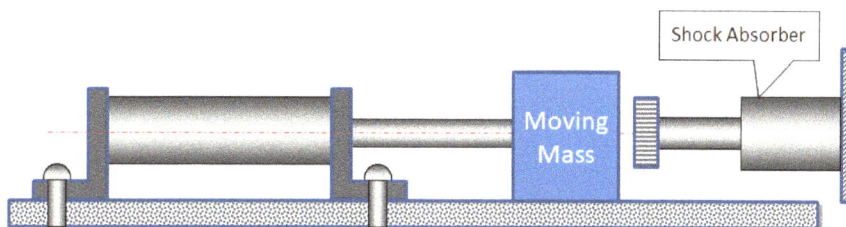

Fig. 6.29 – Protect End Caps Against Impact Load using Shock Absorber

35

6.3.11- Protect Cylinder from External Hazard

- Protecting cylinder body and cylinder rod from external hazard →
- Improves cylinder reliability and operating safely.

❖ **Heat Sources:**
- Overheating →
- pressure intensification in closed oil volumes.
- Improper sealing performance.

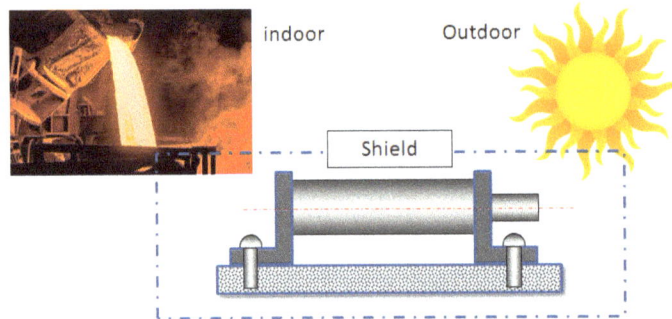

Fig. 6.30 – Shield Hydraulic Cylinders from Heat Source 36

36

❖ **External Hazardous:**
- Examples of hazardous conditions (weld splatter, fast drying chemicals, paint, excessive heat, abrasive contaminants etc.).

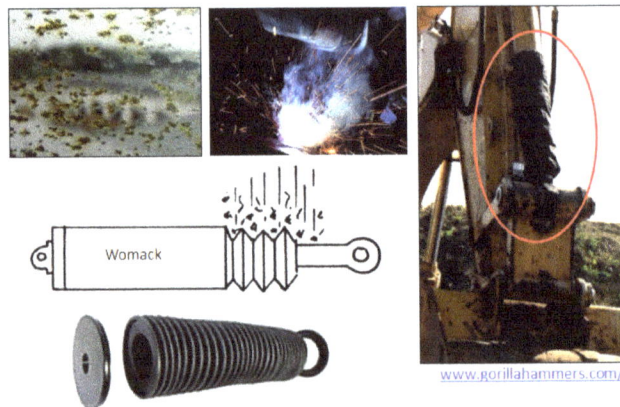

Fig. 6.31 – Shielding Hydraulic Cylinder Rods from Hazardous Conditions 37

37

6.3.12- Protect Cylinder Rod from Corrosion

- Corrosive environment (snow + sea water + high humidity) →
- Rod failure + external leakage.

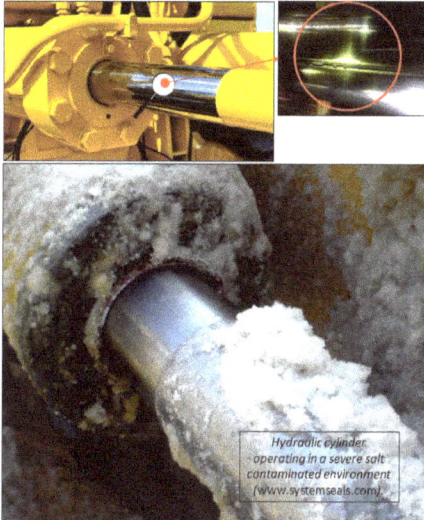

Fig. 6.32 – Protecting Cylinder Rod from Corrosion

38

38

Best Practices:

- Use fresh water to remove salt, sand, or any residues from the piston rod.

- Do NOT use steam cleaners or high-pressure water jets.

- Using industrial lint-free wipe, cover the cylinder rod with protection oil film of low viscosity.

- Apply the protection oil to the entire piston rod.

- Ensure compatibility of the oil used.

- Periodically check the protective oil film depends on how harsh the conditions are.

39

39

6.3.13- Protect Air Chamber of a Single-Acting Cylinder from the Environment

- Single-Acting Cylinder (initially retracted or initially extended) →
- Air breather against dust and humidity is required.
- Air breather physical size and mesh size are specified by the system's designer.

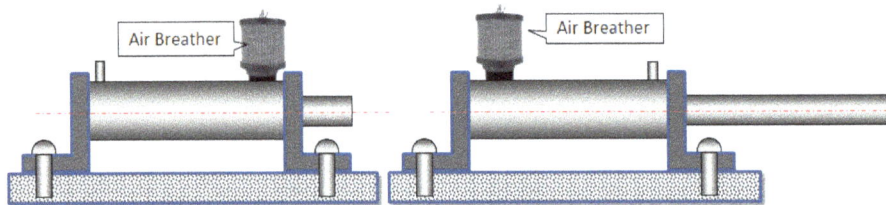

Fig. 6.33 – Air Breather to Protect Air Chamber of a Single-Acting Cylinder

40

40

6.3.14- Review Range of Allowable and Maximum Working Conditions

- Maximum Pressure, temperature, speed, load, etc. are specified by a system manufacturer.
- Maximum pressure in cylinders isn't just set by PRV!!!
- Be aware of pressure intensification in differential cylinder.

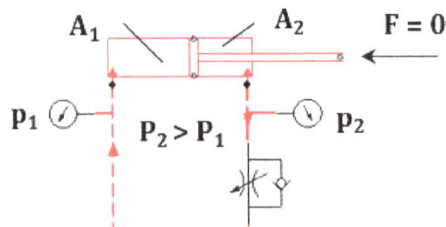

Fig. 6.34 – Pressure Intensification in a Double-Acting Differential Cylinder

Example of Cylinder Maintenance: Video 37 (24 min)

41

41

6.4- BP-Cylinders -04-Standard Tests and Calibration

- **ISO10100-2001** standard specifies the requirements and procedure for cylinder testing.

- This test procedure is designed to check and determine the amount of cross-piston leakage and rod seal leakage.

- It is recommended that the cylinder be tested in both the extend and retract positions.

42

Test Piston Leakage in Retraction Position:
- **Step 1:** Shut the prime mover off.

- **Step 2:** Lock out the electrical system or tag the keylock switch in accordance with local regulations.

- **Step 3:** Observe the system pressure gauge. Release any residual pressure trapped in the system by an accumulator, counterbalance or pilot-operated check valve, suspended load on an actuator, intensifier, or a pressurized reservoir.

- **Step 4:** Install flow meter (2) in series with the transmission line at the rod-end.

- **Step 5:** Install pressure gauge (2) in parallel with the connector at the rod-end.

- **Step 6:** Start the prime mover.

- **Step 7:** Allow the system to warm up to approximately 130 $^\circ$F. (54 $^\circ$C.).

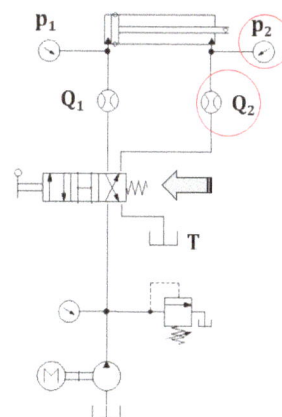

Fig. 6.35 – Hydraulic Cylinder Test Procedure

43

- **Step 8:** Activate the directional control valve to retract the cylinder.

- **Step 9:** While holding in fully retraced position, record on the pressure p_2, flow Q_2, and temperature **T**. If the cylinder isn't leaking past the piston seal in retract, Q_2 will be zero. Q_2 Will only measure leakage. Therefore, its range and resolution must be selected for low flow, i.e. below 4 lit/min (1 gpm). Rather than installing a flowmeter, it would be easier and cost effective is to disconnect the line on the cap end, and just collect the leakage in a bucket using a stopwatch to measure the flow.

- **Step 10:** Release the directional control valve and shut the prime mover off.

- **Step 11:** With the directional valve shifted to extend the cylinder, similarly, repeat step 1 through 10 to test the cylinder leakage in direction of cylinder extension considering the readings p_1, Q_1.

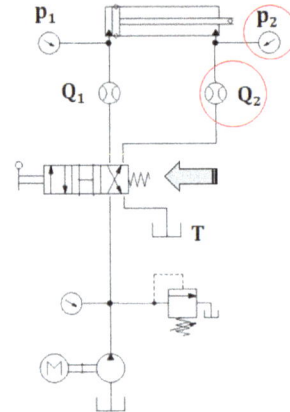

44

44

6.5- BP-Cylinders -05-Transportation and Storage

BP-Cylinders-05- Transportation and Storage (see Fig. 6.36):

- Pack the cylinders in plastic during shipping and transportation.

- Store cylinders indoor in a dry storage area.

- Store cylinders in fully retracted position.

- Store large cylinders vertically → avoid cylinder barrel bending.

- Close cylinder ports with proper dust caps or steel plugs.

- Cover mounting and threaded rod ends with proper protecting covers.

- Long storage → Coat cylinder surface with corrosion preventive.

- Long storage → Fill in with protective fluid specified by manufacturer.

- Long storage → Consider pressure intensification due thermal expansion (Low flow thermal pressure relief valve).

  ΔP = Bulk Modulus x thermal Expansion Coefficient x Temperature Difference
  Example: ΔP (bar) = 20,000 (bar) x 0.0005 (1/$^{\circ}$C) x 50 $^{\circ}$C = 500 bar

45

45

Video 672 (1.5 min)

Fig. 6.36 – Best Practices for Storing Hydraulic Cylinders

46

46

Best Practices for Hydraulic Seals Storage:

❖ **DIN 7716 / BS 3F68: 1977, ISO 2230, or DIN 9088**.
This standard provides fundamental instructions on storage, cleaning and maintenance of elastomeric seal elements.

❖ **Best Practices for Hydraulic Seals Storage:**
1. **General Conditions:** Dry, dust free, and moderately ventilated.

2. **Humidity:** Optimum humidity is 40 to 65 percent, maximum 75%.

3. **Temperature:**
   ▪ Optimum temperature is 25 °C (77 °F),
   ▪ Maximum 50 °C (122 °F).
   ▪ When taken from low temperatures, seals should be warmed up to approximately 30°C (86ºF) before they are used.
   ▪ Warming them up shouldn't be fast, an hour of time span is fine.

4. **Air:** Avoid exposure to direct and continuous stream of conditioned air.

47

47

5. **Heat:** Avoid exposure to direct heat source such as boilers or radiators.

6. **Light:**
   - Avoid exposure to direct sunlight and ultraviolet light.
   - Pack in opaque containers, it is advisable to
   - Cover windows with red or orange screens or coatings.

7. **Radiation:** Avoid exposure to Gamma radiation, otherwise seal compression set is severely affected.

8. **Ozone and Oxygen:**
   - Avoid exposure to sources of ozone such as mercury vapor lamps, high-voltage electrical equipment, combustible gases, and organic vapors.
   - Wrap in airtight containers.

9. **Liquids:** Avoid exposure to gasoline vapors, greases, acids, solvents.

10. **Contact with Elastomers:** Avoid contact between seals made from dissimilar compounds. Each type of seals is backed individually.

48

11. **Contact with Metals:** Avoid contact with certain metals that have degrading effects on some elastomers such as manganese, iron and particularly copper.

12. **Packaging:** Pack the seals in stress-free cases and DO NOT squeeze a hydraulic seal to accommodate it in a small storage area.

13. **Deformation:**
    - DO NOT store seals on top of each other
    - DO NOT place heavy objects on top of any stored seals.
    - Store seals in a relaxed position, free from tension or compression.

14. **Hanging:** DO NOT hang or suspend seals on a hook in a vertical position as gravity will distort the seal over time.

15. **Cleaning:**
    - Use cleaning fluid that is only specified by seal manufacturer.
    - Organic solvents (trichloroethylene, carbon tetrachloride, and petroleum are the most harmful agents.
    - Soap, water, and methylated spirits are the least harmful.
    - All parts should be dried at room temperature before use.

49

16. **Stock Rotation (FIFO):** Stock the seals in rotation, i.e. First-In, First-Out manner (FIFO). This ensures that the next seal used in the rotation will be within its intended shelf life.

17. **Shelf Life:** In 1998, SAE issued an Aerospace Recommended Practice (ARP) for the storage time of elastomer seals and seal assemblies prior to installation.

50

Compound Name	ASTM Polymer	Shelf Life
Aflas®	FEPM	Unlimited
Butyl Rubber, Isobutylene Isoprene	IIR	Unlimited
Chloroprene (Neoprene®)	CR	15 Years
Chlorosulphonated Polyethylene (Hypalon®)	CSM	15 Years
Epichlorohydrin (Hydrin®)	ECO	NA
Ethylene Acrylic (Vamac®)	AEM	15 Years
Ethelene Propylene, EPDM or LP	EP	Unlimited
Fluorocarbon (Viton®)	FKM	Unlimited
Fluorosilicone	FVMQ	Unlimited
Hydrogenated Nitrile, HNBR or HSN	HNBR	15 Years
Nitrile (BUNA-N or NBR)	NBR	15 Years
Perfluoroelastomer	FFKM	Unlimited
Polyacrylate	ACM	15 Years
Polyurethane (Polyester or Polyether)	AU/EU	5 Years
Silicone	Q,VMQ,PVMQ	Unlimited
Styrene Butadiene (Buna-S)	SBR	3 Years

Table 6.2 - Approximate Shelf Life for Standard Elastomers (Courtesy of MFP Seals)

51

Best Practices for Hydraulic Cylinders Lifting:

❖ Improper lifting → minor-to-major damage to the

❖ Review instructions by manufacturer.

❖ Best Practices for Cylinder Lifting:

▪ DO NOT Lift a cylinder in when the it is in vertical position and the cylinder rod faces the ground. This may result in unexpected cylinder extension during lifting.

▪ DO NOT hang the cylinder from the rod and the barrel simultaneously because this may cause rod seal damage. It rather recommended to hang the cylinder from the barrel only with the hanging attachments are equidistant from the cylinder's center of gravity.

▪ DO NOT lift the cylinder while it is filled with oil.

52

52

Chapter 6 Reviews

1. 1. Cylinder rod misalignment with the attached load may lead to?
    A. Increased cylinder speed.
    B. Cylinder creeps or load drifts.
    C. Destruction of rod bearing, rod buckling and reduced service life of the cylinder.
    D. Reduced cylinder stroke.

2. It is recommended to equip the cylinder by internal cushioning devices when the cylinder speed exceeds?
    A. 4 cm/s.
    B. 6 cm/s.
    C. 8 cm/s.
    D. 10 cm/s.

3. If a smaller cylinder is used to replace a defective larger cylinder, assuming the supply flow and loading conditions are the same, what would be the expected changes in the system performance?
    A. Increased cylinder speed and increased working pressure.
    B. Increased cylinder speed and decreased working pressure.
    C. Decreased cylinder speed and increased working pressure.
    D. Decreased cylinder speed and decreased working pressure.

4. Look at how the devices shown below are installed and choose the correct statement?
    A. Only cylinder limit switch is installed correctly.
    B. Only directional valve installed correctly.
    C. Only hydraulic conductors installed correctly.
    D. None of the shown devices are installed correctly.

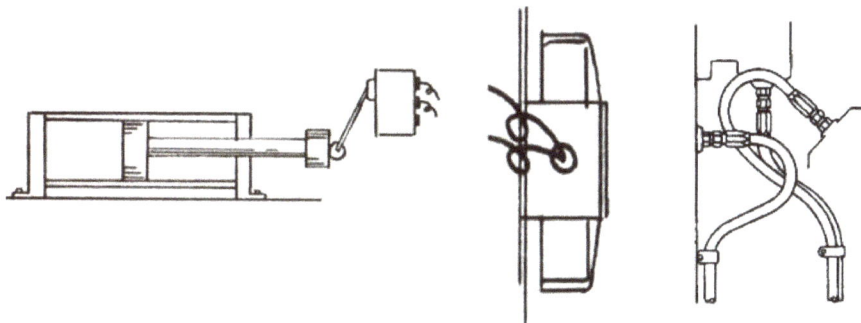

5. When lifting a hydraulic cylinder, which of the following statements is considered "TRUE"?
    A. DO NOT Lift a cylinder in when the it is in vertical position and the cylinder rod faces the ground.
    B. Hang the cylinder from the rod and the barrel simultaneously
    C. No problem lifting the cylinder while it is filled with oil.
    D. None of the above statements are TRUE.

Chapter 6 Assignment

Student Name: --- Student ID: ------------------

Date: -- Score: ------------------------

Question: Explain why air must be bled out of a hydraulic cylinder before first use or after major maintenance.

**Chapter 7
Maintenance of Valves**

Objectives:

This chapter provides guidelines for **valves** selection, replacement, maintenance scheduling, installation, testing, storage and transportation. This chapter is supported by examples and figures granted by leading fluid power manufacturers.

Brief Contents:

0

0

7.1- BP-Valves-01-Selection and Replacement

Valves are originally specified to consider:

- Correct function.
- Leakage control.
- Maintenance and adjustment requirements.
- Resistance against environmental influence.
- This section is not intended to provide design solutions or valve sizing.

When selecting or replacing an existed cylinder →

❖ **Valve Function:**

- The new valve must be identical symbol and part number.
- Minor difference → major functional difference.

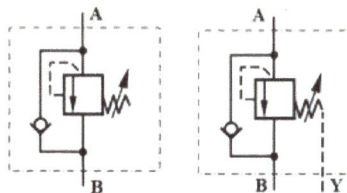

Fig. 7.1- Symbols for Counterbalance Valve (Left) and Sequence Valve (Right)

1

1

❖ **Directional Valve Transitional Conditions:**

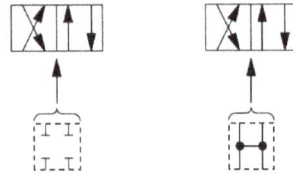

Fig. 7.2- Different Transitional Conditions for a Directional Valve

❖ **Valve Size:**
- Valves are primarily sized based on the flow rate.
- Oversized valve → loses controllability.
- Undersized valve → performs like a throttle valve generating heat.

❖ **Valve Operating Conditions:**
- Review maximum (P, T, Cleanliness requirements, etc.)
- Example: Upgrading an On/Off control system to continuous control mode using a proportional valve requires new filtration system.

❖ **Electrical Components on EH Valves:**
- Review electrical design parameters (voltage, current, power, switching time, cyclic rate, etc.)

2

2

7.2- BP-Valves-02-Maintenance Scheduling

Video 177 (0.5 min)

Unless otherwise stated by components and systems manufacturer:

#	Preventive Maintenance Actions	Daily	Weekly	Monthly	Biannually	Annually
1	Clean around and outside surface	✔	✔	✔	✔	✔
2	Check for proper connection with the hydraulic lines.		✔	✔	✔	✔
3	Check for proper connection with the electrical lines (if found)		✔	✔	✔	✔
4	Valve major maintenance and testing.				✔	✔

Table 7.1- BP-Valves-02-Maintenance Scheduling

3

3

7.3- BP-Valves-03-Installation and Maintenance

❖ Hydraulic valves must be kept well-maintained to assure proper system functionality.

❖ Standards <u>ISO/DIS 4411</u> **and** <u>ISO 4411:2019</u> provide detailed information about valve installation and adjustment.

❖ **BP-Valves-03-Installation and Maintenance:**
1. Proper Valve Disassembly.
2. Proper Valve Inspection and Maintenance.
3. Proper Valve Assembly.
4. Proper Hydraulic Connections
5. Proper Electrical Connections
6. Proper Valve Adjustment.

4

4

7.3.1- Proper Valve Disassembly

Hydraulic Valves are very sensitive to contamination →
▪ keep the surrounding area clean is highly recommended.

▪ **Cleaning Towels (1):** Industry specified and lint-free.

▪ **Protective Plates and Dust Caps (2):** Are removed only immediately prior to installation.

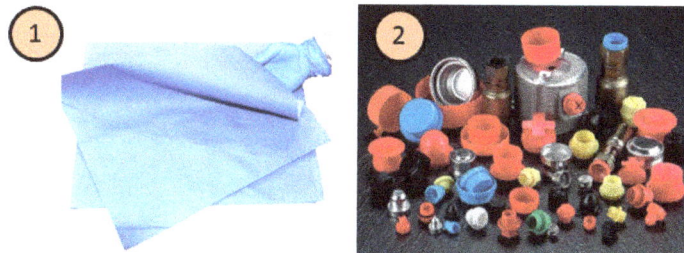

Fig. 7.3- Proper Housekeeping when disassembling a Hydraulic Valve

5

5

7.3.2- Proper Valve Inspection and Maintenance

Clean Outside Surfaces of the Valve:

Metallic chips are attracted by the electrical parts + Dirt accumulated on the outside surfaces →

- Dirt will find a way to get inside.
- Reduced heat dissipation from the outside surfaces of the valve.

Fig. 7.4- Improper Maintenance Practices Kept the Valve Dirty

6

6

Last chance Filter Inspection: in servo valves and enhanced performance proportional valves, there must be a miniature filter installed usually in the inlet pressure and called *last chance filter*. It is mandatory to inspect and clean this filter during maintenance.

Fig. 7.5- Last Chance Filters in Servo Valves

7

7

7.3.3- Proper Valve Assembly

- **Gravity (1):**
- **Line-Mounted Valves (2):**
- **Actuator-Mounted Valves (3):**
- **Surface-Mounted Valves (4):**
- Whether using a subplate (as in
- **Flange-Mounted Valves (5):** When welding a flange to a pipe, remove the flange seal until the flange has cooled down then replace the seal.

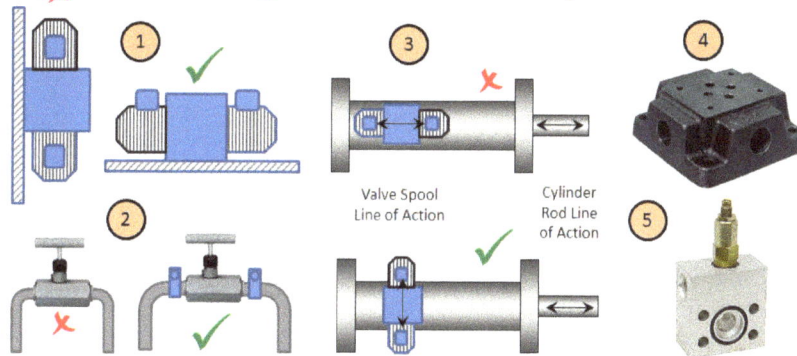

Fig. 7.6- Best Practices for Proper Valve Mounting

8

8

- **Mounting Torque:** Bolting torque mustn't exceed manufacturer specifications in order to avoid valve body distortion. Clearances inside valves are very small and over torqueing may result in spool seizure.

- **Space Surrounding:**
- Enough clearance for wrench and/or bolt access and electrical connections.
- Access for removal, repair or adjustment.

- **Components Surrounding:** Consider minimizing the chance of damaging the valve by surrounding mechanical operating devices.

- **Electrical Connections:** Shall be in accordance with appropriate standards (e.g. IEC 60204-1 or manufacturer standard) and be designed with the suitable protection class (e.g. in accordance with IEC 60529).

9

9

Example of Manufacturer's Instructions:

Step 1: Remove cartridge from packing. Inspect O-Ring to ensure there is no damage such as cuts or nicks. Check if the backup rings fit tightly within the O-ring groove.

Step 2: Immerse the hydraulic portion of the cartridge valve in clean compatible oil to lubricate the seals. Dry seals could cause the backup rings to spin out of the groove which damage or cut the seal.

Step 3: Insert the cartridge valve into the cavity and tighten by hand in a clockwise manner. You should be able to screw it in with little resistance up to the O-ring.

Step 4: Continue to screw in the cartridge with a torque wrench and tighten to the torque specified in the catalog. Tightening the valve above the specified torque value, this may cause the spool or poppet to stick.

Step 5: Install the waterproof O-ring on the cartridge hex if one is required.

Step 6: Install the coil with the lettering facing the hex nut. Install the coil nut and tighten the coil nut to specified torque.

10

10

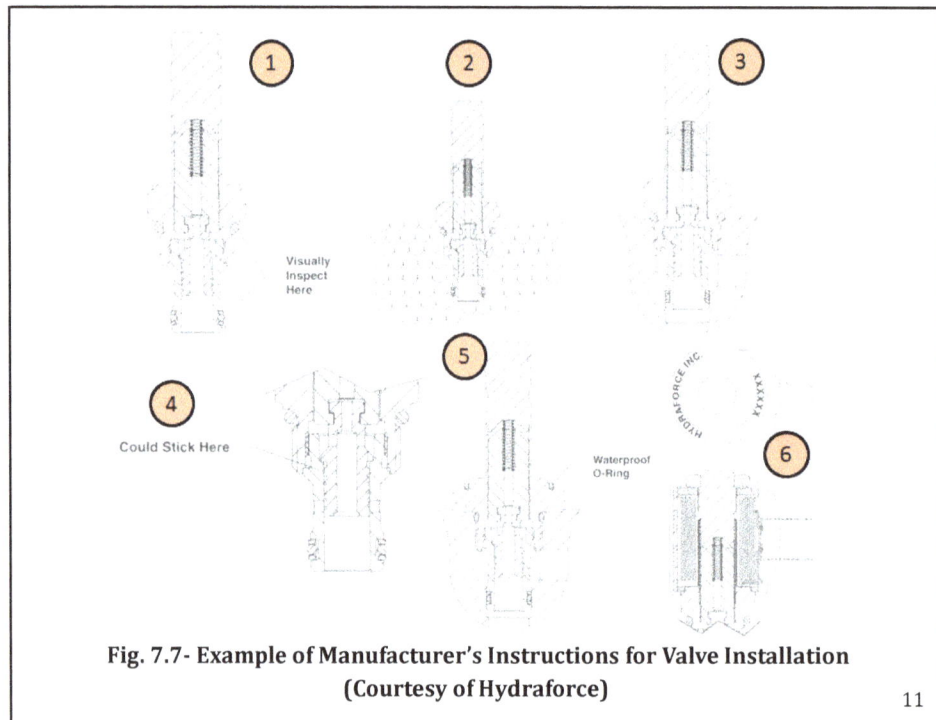

**Fig. 7.7- Example of Manufacturer's Instructions for Valve Installation
(Courtesy of Hydraforce)**

11

7.3.4- Proper Hydraulic Connections

Example 1 – Main Line Connection:
If the load lines aren't connected to the right sides of the actuator →
the actuator motion will be reversed.

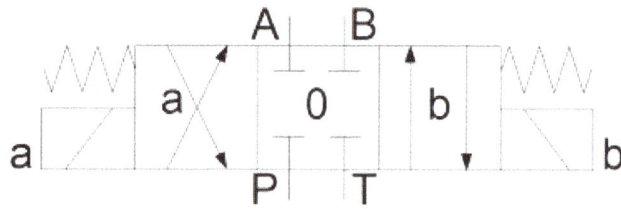

Fig. 7.8- Main Ports on a 4-way Directional Valve

12

12

Example 2 – Pilot Line Connection:
- Incorrect supply and drain of pilot pressure → valve malfunction.
- A pressure filter without by-pass is required immediately before the valve in the control pressure line to port "X" ($\beta10 > 75$ to ISO OR cleanliness class 5 to NAS 1638).

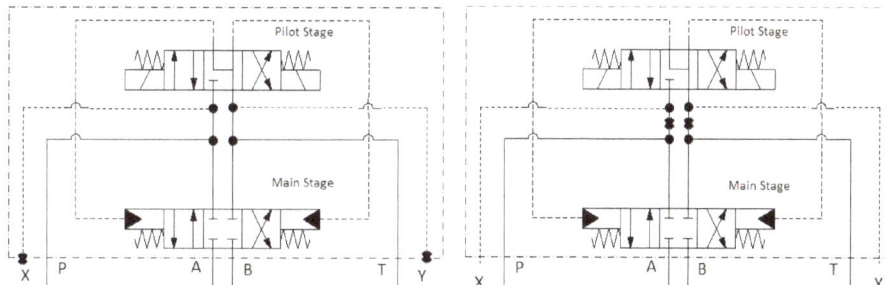

Fig. 7.9- Supply and Drain of Pilot Pressure in a Pilot-Operated DCV

13

13

7.3.5- Proper Electrical Connections

- **Plug-in Connectors (1):** no flying wires to avoid short circuiting.
- **Plug with Built-in light Indicator (2):**
- **Keep Solenoids Covered (3):**
- **Match the Solenoids with the Actuator's Motion (4):**
- **Electrical Cables:** Keep it away from plumbing or source of noise

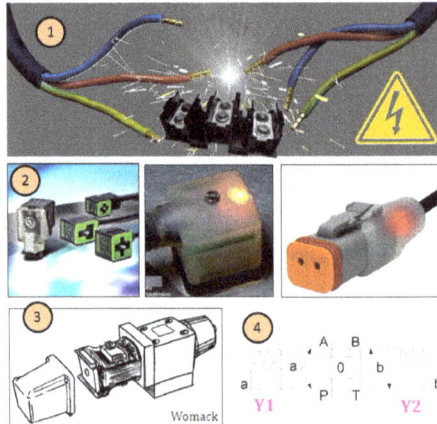

Fig. 7.10- Considerations for Proper Electrical Connection to Hydraulic Valves

14

14

7.3.6- Proper Valve Adjustment

- Improper adjustment → potential hazard.

- Improper adjustment of FCV → improper actuator speed.

- Improper adjustment of DCV → improper actuator Direction.

- Improper adjustment of PCV → improper load carrying capacity.

- When valves permit adjustments of one or more parameters, the following provisions should be incorporated, as appropriate:

  - Means for securing the adjustment.
  - Means for locking the adjustment, if required to prevent unauthorized change; or
  - Means for preventing adjustment beyond a safe range.

15

15

- Valve adjustment → permitted and experienced persons.

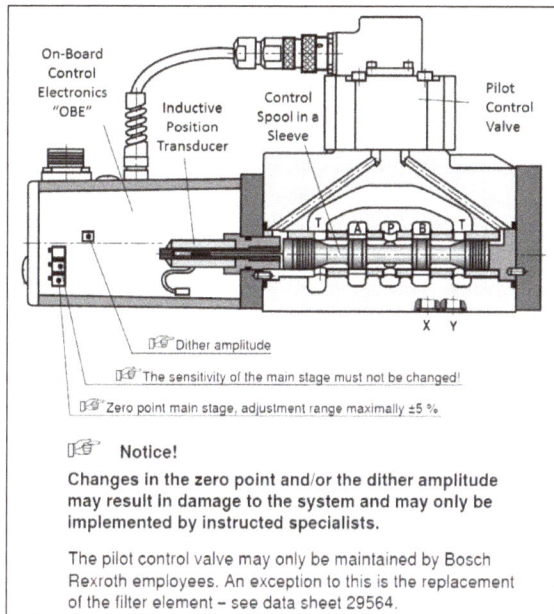

☞ Dither amplitude

☞ The sensitivity of the main stage must not be changed!

☞ Zero point main stage, adjustment range maximally ±5 %

☞ **Notice!**

Changes in the zero point and/or the dither amplitude may result in damage to the system and may only be implemented by instructed specialists.

The pilot control valve may only be maintained by Bosch Rexroth employees. An exception to this is the replacement of the filter element – see data sheet 29564.

Fig. 7.11- Instructions for Servo Valve Adjustments (Courtesy of Bosch Rexroth)

16

16

7.4- BP-Valves-04-Standard Tests and Calibration

This section was graciously granted to this textbook as a Courtesy of "CFC Industrial Training".

CAUTION! The pressure and flow ratings of the diagnostic equipment which will be used to conduct these tests must be at least equal, if not greater than, the pressure and flow ratings of the valve being tested.

Test Conditions: To get comparable and unified test results, tests must be conducted:
- At a predefined temperature (e.g. approximately 130 °F (54 °C.)
- Using hydraulic fluid that has a predefined viscosity (e.g. 32 cSt.).

17

17

7.4.1- Flow Control Valve Test

Purposes of the Test:
- Determine the leakage across the valve when it is in the "closed" position.

Test Procedure (Figure 7.12):
- Using a hand pump (1), push the oil to the inlet port of the needle valve.
- Observe pressure gauge (2), it should "hold" for a reasonable time.
- There is no general rule-of-thumb regarding leakage rate.
- Refer to the valve manufacturer's specifications.
- Open the pressure relief valve and release the pressure.

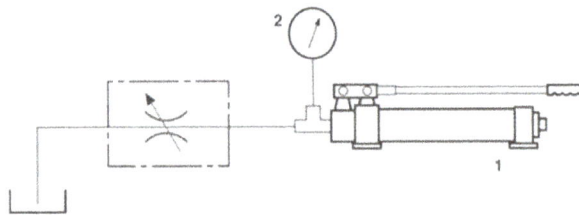

Fig. 7.12- Standard Test Circuit for Flow Control Valve (CFC Industrial Training) 18

18

7.4.2- Test for Pressure Relief Valves

Note: This test is also applicable for counterbalance and sequence valves.

Test A: check if the valve responds to pressure adjustment:
- Keep the actuator in deadheaded "stalled" position.
- Keep the PRV fully opened.
- Keep the needle valve (3) fully closed (turn clockwise).
- Turn on the pump and observe pressure gauge (1), it should read tank line.
- Close the PRV gradually to maximum pressure → **p** increases accordingly.

Test B: develop the valve characteristic curve:
- Keep the actuator in deadheaded "stalled" position.
- Keep the PRV in its required adjustment.
- Keep the needle valve (3) fully opened (turn counterclockwise).
- Turn on the pump and observe pressure gauge (1), it should read tank line.
- Gradually and incrementally, based on the pressure, close the needle valve (3).
- Plot the flow "**Q**" at the flow meter (4) versus the pressure "**p**" at the pressure gauge (1).

19

19

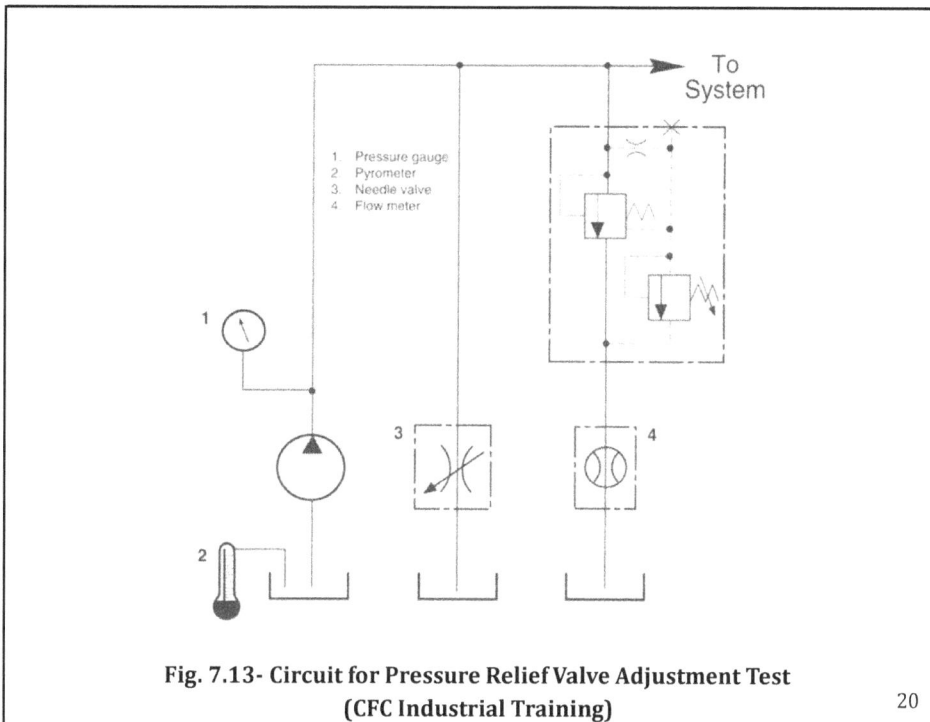

1. Pressure gauge
2. Pyrometer
3. Needle valve
4. Flow meter

**Fig. 7.13- Circuit for Pressure Relief Valve Adjustment Test
(CFC Industrial Training)**

20

20

Test C: Determine the leakage across the valve:

- Using a hand pump (1), push the oil to the inlet port of the needle valve.
- Observe pressure gauge (2), it should "hold" for a reasonable time.
- There is no general rule-of-thumb regarding leakage rate.
- Refer to the valve manufacturer's specifications.
- Open the pressure relief valve and release the pressure.

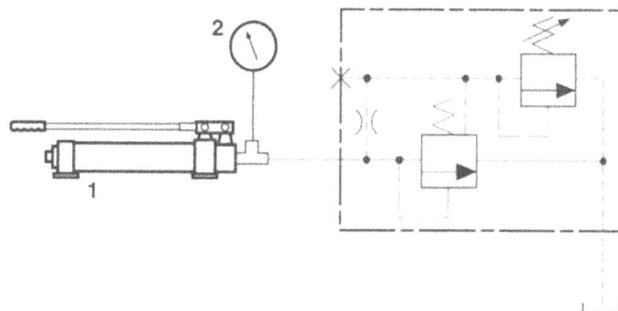

**Fig. 7.14- Circuit for Pressure Relief Valve Leakage Test
(CFC Industrial Training)**

21

21

7.4.3- Test for Pressure Reducing Valves

Purposes of the Test:

- Check the operation of the valve.
- The pressure in the external drain-line.
- If there is a restriction in the external drain-line.
- If there are pressure surges in the external drain-line.

Test Procedure (Figure 7.15):

- Open needle valve fully (turn counterclockwise).
- Turn on the pump.
- Gradually close the needle valve and observe the pressure gauges 1,2, & 3.
- Gauge 3 supposed to read approximately constant reduced pressure (p_3) until the valve is in the inactive zone, where the differential pressure (p_2-p_3) is reduced below the value specified by the valve manufacturer.
- Gauge 2 shows if there is blockage or surge leakage in the spring chamber vent line.

22

22

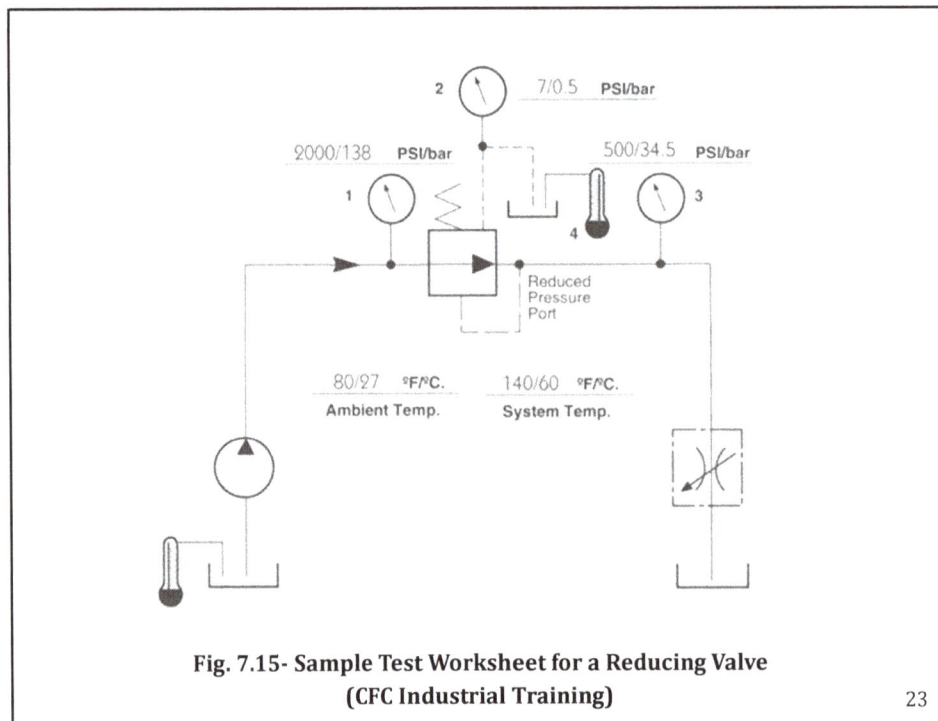

Fig. 7.15- Sample Test Worksheet for a Reducing Valve (CFC Industrial Training)

23

23

7.4.4- Directional Control Valve Test

Purposes of the Test: Determine the leakage across the valve.

Test Procedure:
- Block the port on the leakage path under the test. In this figure P-B.
- Pressurize this pass to predefined maximum pressure of the valve.
- Observe pressure at the gauge.
- The pressure should "hold" for a reasonable length of time.
- Open the pressure relief valve and release the pressure.

24

24

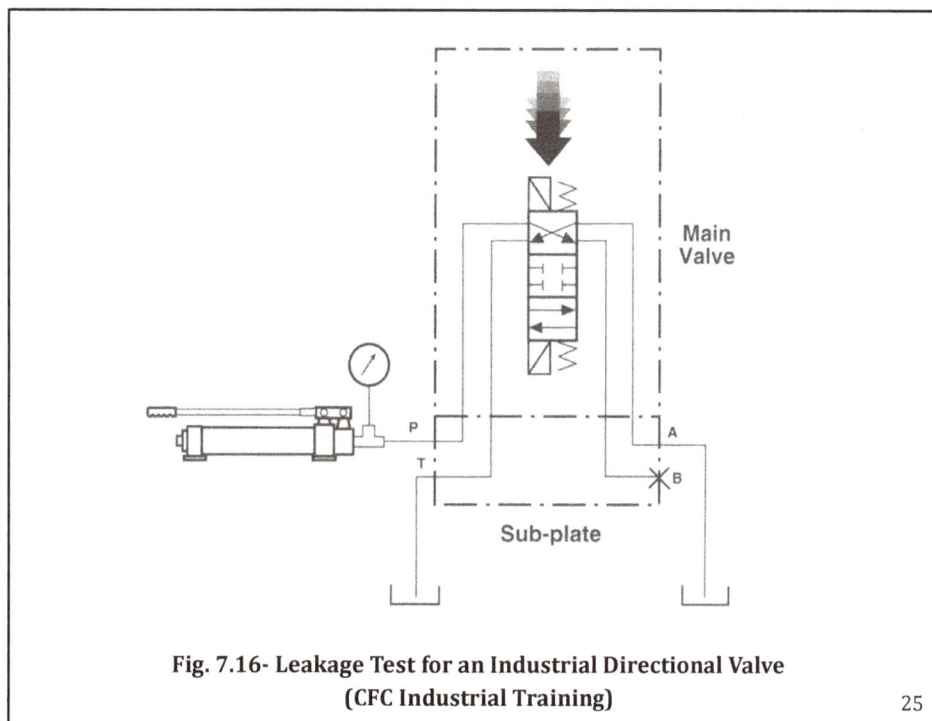

**Fig. 7.16- Leakage Test for an Industrial Directional Valve
(CFC Industrial Training)**

25

25

7.4.5- Proportional and Servo Valve Test

Regulating Standard (ISO 10770-1-2009): Describes methods for determining the performance characteristics of electrically modulated, hydraulic, four-port directional flow-control valves. This type of electrohydraulic valve controls the direction and flow in a hydraulic system.

Type of Tests: There are number of standard tests that can be conducted to check the performance of EH valves. These tests are conducted if required to diagnose some malfunctions of the valve. It can also be conducted frequently as part of preventive maintenance plans. The following are the standard tests:

- Static Test: Flow Gain or Pressure Gain.

- Static Test: Limiting Power.

- Static Test: Hysteresis.

- Static Test: Null leakage and Proof Pressure.

- Dynamic Test: Step response.

- Dynamic Test: Frequency response.

- Dynamic Test: Fail-Safe Function.

26

26

Standard Test Conditions: These test conditions must be maintained to receive comparable results.

Parameter	Condition
Ambient temperature	20 °C ± 5 °C
Fluid cleanliness	Solid contaminant code number shall be stated in accordance with ISO 4406.
Fluid type	Commercially available mineral-based hydraulic fluid (i.e. L - HL in accordance with ISO 6743-4 or other fluid with which the valve is able to operate)
Fluid viscosity	32 cSt ± 8 cSt at valve inlet
Viscosity grade	Grade VG32 or VG46 in accordance with ISO 3448
Pressure drop	Test requirement ± 2.0 %
Return pressure	Shall conform to the manufacturer's recommendations

**Table 7.2- Standard Test Conditions for Four-Ports EH Valve
(Courtesy of NFPA)**

27

27

Standard Test Circuit: The standard test circuit that contains components for (testing, condition monitoring, and outputting the test results).

Caution: A valve MUST NOT be tested with electrical signal only before hydraulic power is supplied including pilot pressure to the first stage. Otherwise, the wire spring in a flapper nozzle stage may be severely stretched. Same is applicable for proportional valves in order to prevent spool friction without lubrication.

#	Component	#	Component
1	Main Flow Source	14	Pressure Gauge
2	Main Relief Valve	15	Signal Conditioner
3	External Pilot Flow Source	16	Data Acquisition
4	External Pilot Relief Valve	S1 – S9	Shut-off Valves
5	Unit under Test	A, B	Load Ports
6 - 9	Pressure Transducer	P	Supply Pressure Port
10, 11	Flow Transducers	T	Tank Port
12	Signal Generator	X	Pilot Control Pressure Supply
13	Temperature Indicator	Y	Pilot Control Pressure Return

Fig. 7.17- Standard Test Circuit for Four-Ports EH Valve (Courtesy of NFPA) 28

28

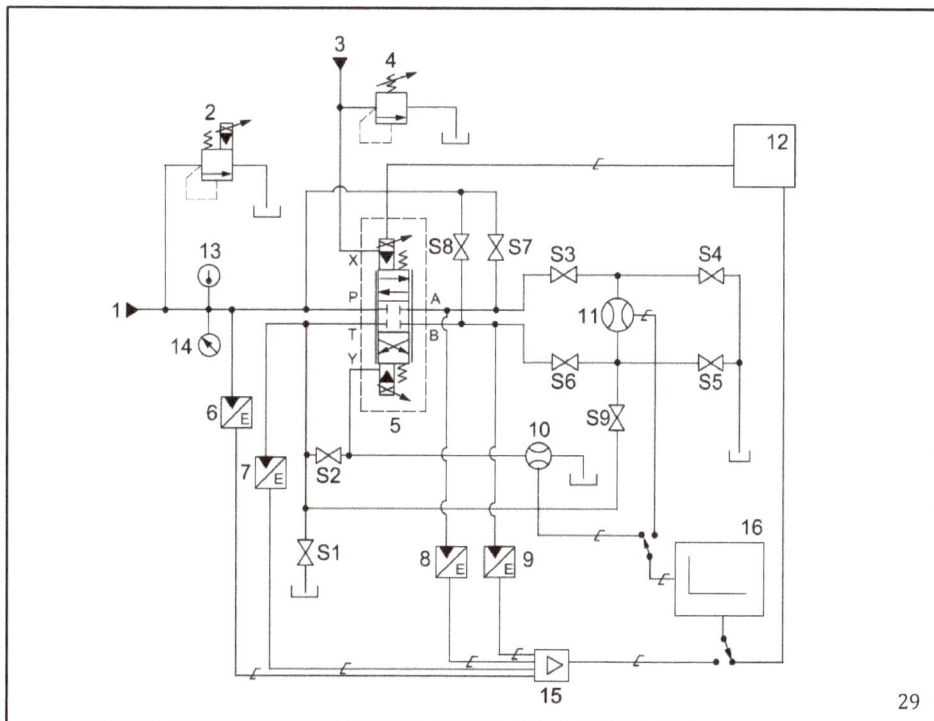

29

Test Accuracy: Instrumentation shall be accurate to within the limits shown in **Class B of ISO 9110-1:**

- Electrical resistance: ± 2 % of the actual measurement.
- Pressure: ± 1 % of the valve's rated pressure drop to achieve rated flow.
- Temperature: ± 2 % of the ambient temperature.
- Flow: ± 2,5 % of the valve's rated flow.
- Input signal: ± 1,5 % of the electrical input signal required to achieve the rated flow.

Commercial Servo and Proportional Valve Testers: the following couple of examples shows typical commercial proportional valve testers available in the market.

30

30

Example 1: Stationary Valve Tester: The test rig is integrated with a hydraulic power supply. The features of this test rig in brief are as follows:
- Used for both manual and automatic tests.
- Used for both steady state and dynamic performance tests.
- Plug and Play test procedure, no high skills required.
- Suitable for various valve electrical input signals.

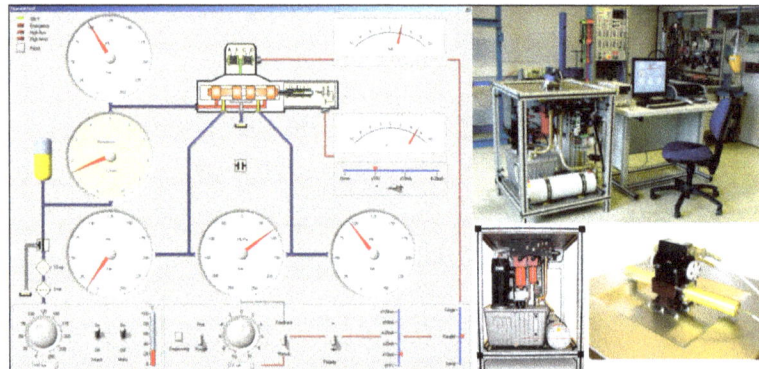

Fig. 7.18- Stationary Proportional and Servo Valve Tester
(dietzautomation.com)

31

31

Example 2: Portable Valve Tester:

Fig. 7.19- Portable Servo Valve Analyzer Series F087-127 (Courtesy of Moog)

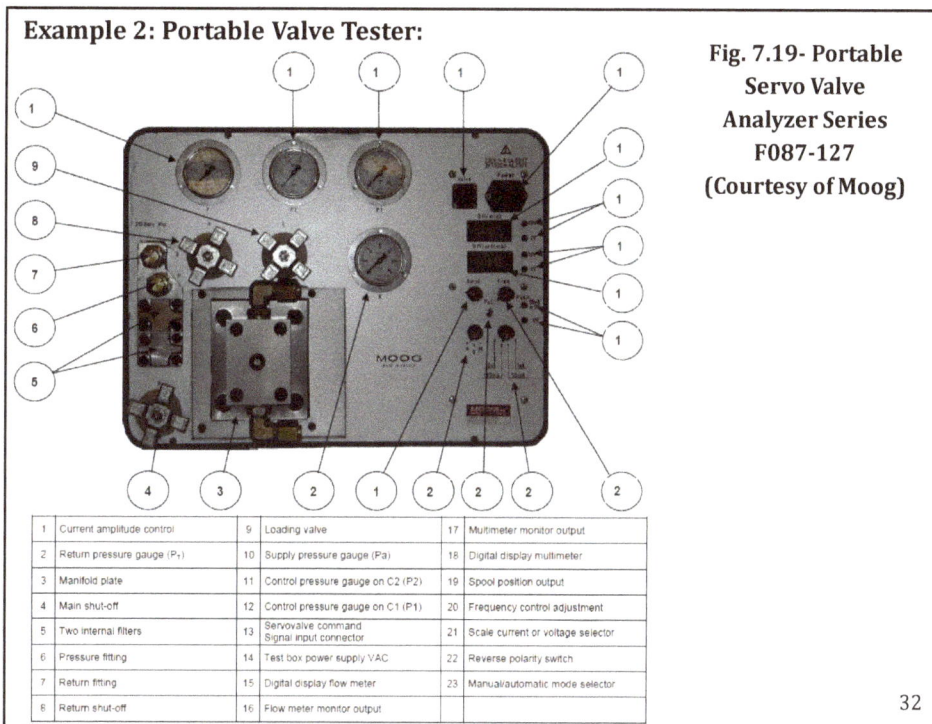

1	Current amplitude control	9	Loading valve	17	Multimeter monitor output
2	Return pressure gauge (P₁)	10	Supply pressure gauge (Pa)	18	Digital display multimeter
3	Manifold plate	11	Control pressure gauge on C2 (P2)	19	Spool position output
4	Main shut-off	12	Control pressure gauge on C1 (P1)	20	Frequency control adjustment
5	Two internal filters	13	Servovalve command Signal input connector	21	Scale current or voltage selector
6	Pressure fitting	14	Test box power supply VAC	22	Reverse polarity switch
7	Return fitting	15	Digital display flow meter	23	Manual/automatic mode selector
8	Return shut-off	16	Flow meter monitor output		

32

32

7.5- BP-Valves-05-Transportation and Storage

- The storage locations → dry + low humidity + dirt-free.

- Storage longer than 3 months:
  - → It is always advisable to refer to the manufacturer's specifications for longer term storage.

  - → Fill the housing with compatible preservative oil and seal the valve with protective covers.

  - → The stored valves must be checked from time to time to make sure that they are kept in the right position.

33

33

Example of Valve Maintenance

Video 43-On-Off Valve (8.5 min)

Video 393-On/Off Valve (8.5 min)

Video 673-On/Off Valve (10 min)

Video 46-Pilot Operated Valve (22 min)

Video 50-Flow Control Valve (11 min)

Video 86-Servo Valve (10.5 min)

Video 86-Proportional Valve (17 min)

34

34

Chapter 7 Reviews

1. The symbol shown below is for a pilot-operated (2-stage) directional control valve. The central position of the pilot stage of this valve is of a?
    A. Float center.
    B. Tandem center.
    C. Closed center.
    D. Open center.

P T

2. The symbol shown below is for a pilot-operated (2-stage) directional control valve. The central position of the main stage of this valve is a?
    A. Float center.
    B. Tandem center.
    C. Closed center.
    D. Open center.

P T

3. The symbol shown below is for a pilot-operated (2-stage) directional control valve. By energizing any one of the pilot stage's solenoids, the actuator attached to the ports A and B does not move. The reason is?
    A. Internal supply of the pilot pressure escapes from an open-center main stage's spool.
    B. External supply of the pilot pressure escapes from an open-center main stage's spool.
    C. External supply of the pilot pressure escapes from a tandem-center main stage's spool.
    D. Internal supply of the pilot pressure escapes from a tandem-center main stage's spool.

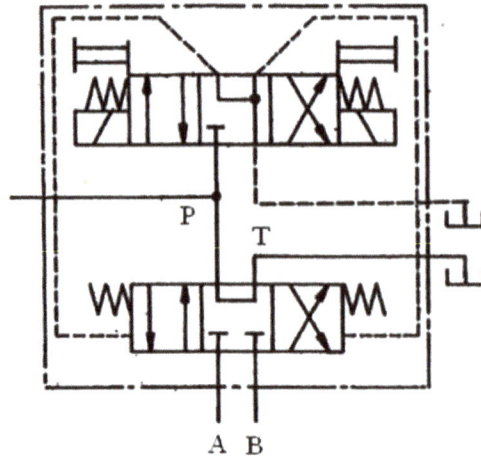

4. To resolve the issue shown in the previous problem, which of the following is the right action?
   A. Place a spring-loaded check valve at the entrance of the main spool stage.
   B. Use external supply of control pressure.
   C. Use closed-center main spool if possible.
   D. All the above statements are correct.

5. The circuit shown below is made to prevent energizing solenoids Y1 and Y2 at the same time in order to avoid burning the solenoids. There is one contactor was drawn incorrectly. Select the correct action to resolve the issue?
   A. Contactor K1 (32-31) must be normally closed.
   B. Contactor K2 (32-31) must be normally opened.
   C. Contactor K1 (11-14) must be normally closed.
   D. Contactor K1 (11-14) must be normally closed.

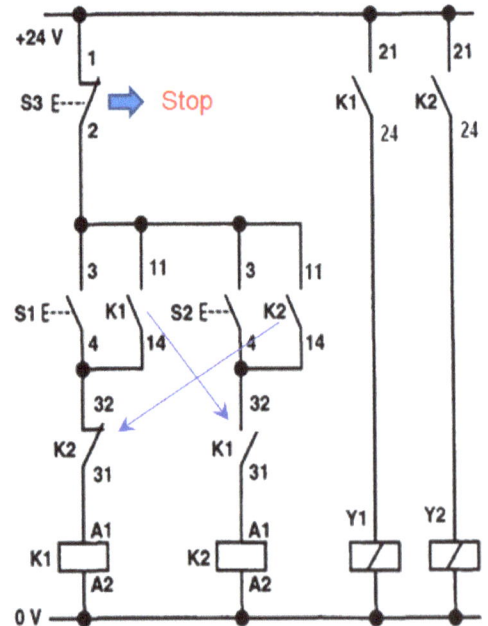

Chapter 7 Assignment

Student Name: -- Student ID: ------------------

Date: -- Score: ------------------------

Questions: List the types of tests to check on performance of proportional or servo valves. Choose one of the tests and explain the outcome and the importance of this test.

Chapter 8
Maintenance of Accumulators

Objectives:

This chapter provides guidelines for **accumulator's** selection, replacement, maintenance scheduling, installation, testing, storage and transportation. This chapter is supported by examples and figures granted by leading fluid power manufacturers.

Brief Contents:

8.1-BP-Accumulators-01-Selection and Replacement

8.2-BP-Accumulators-02-Maintenance Scheduling

8.3-BP-Accumulators-03-Installation and Maintenance

8.4-BP-Accumulators-04-Standard Tests and Calibration

8.5-BP-Accumulators-05-Transportation and Storage

0

0

8.1- BP-Accumulators-01-Selection and Replacement

Accumulators are originally specified to

- Work safely.

- Store the required amount of energy (volume of oil under pressure)

- Respond to the system needs on steady and dynamic modes.

- This section is not intended to provide design solutions or accumulator sizing (For these purposes review volume 1).

When selecting or replacing an existed cylinder →

- **Nominal Size and Initial Gas Pressure:** Nominal size and initial gas pressure of the accumulator shouldn't be changed. Otherwise, accumulator may not be function correctly due to the change in the stored energy.

1

1

Information Marked on the Accumulators: Make sure the following information is marked permanently on the body of the vessel.

- Manufacturer's name, logo, serial number, and date of manufacture (month/year).
- Total shell volume (*Nominal Volume*), expressed in liters or gallons.
- Allowable operating temperature range expressed in $^{\circ}$C or $^{\circ}$F.
- Permissible maximum pressure expressed in megapascals or bar or psi.

Note 1:

- The place and method of stamping shall not reduce strength of the shell.
- If space is not available, $\rightarrow$ use tags that are permanently attached to the accumulators.

2

2

8.2- BP-Accumulators-02-Maintenance Scheduling

Unless otherwise is stated by components and systems manufacturer, Table 8.1 provides guidelines for *scheduling* preventive maintenance actions for hydraulic accumulators.

#	Preventive Maintenance Actions	Daily	Weekly	Monthly	Biannually	Annually
1	Check line connections			✔	✔	✔
2	Check mounting components			✔	✔	✔
3	Check gas precharge pressure	The gas precharge pressure should be checked at least once during the first week of operation. If there is no loss of gas precharge pressure, it should be rechecked in 3 to 4 months. Thereafter, it should be checked at least once a year.				

Table 8.1- BP-Accumulators-02-Maintenance Scheduling

3

8.3- BP-Accumulators -03-Installation and Maintenance

- Improper treatment of energy storage components → Hazard of explosion.

- **ISO 5596** Hydraulic Fluid Power - Gas-loaded accumulators with separator Ranges of pressures and volumes and characteristic quantities.

- **ISO 10945** Hydraulic Fluid Power - Gas-loaded accumulators - Dimensions of gas ports.

- **ISO 10946** Hydraulic Fluid Power - Gas-loaded accumulators with separator selection of preferred hydraulic ports.

- Standard **NFPA/T3.4.7 R2-2000 (R2014)** provides detailed information about accumulator's installation and adjustment.

4

4

BP-Accumulators-03-Installation and Maintenance:

1. Review Disassembly Instructions.

2. Review Inspection Instructions.

3. Review Assembly Instructions.

4. Review Charging Instructions.

5

5

8.3.1- Disassembly Instructions

- **Safety:** Accumulators are inherently dangerous due to high pressure gases and fluids.

- **DO NOT ATTEMPT** to maintain these systems unless:
  o Adequately trained.
  o Wear appropriate safety equipment.
  o Read and understand all instructions provided by the manufacturer.

- **Discharge Fluid:**
  o Free release of the fluid → fast gas expanding → flow surge.
  o Proper discharge rate → review manufacturer's rating.

- **Discharge Gas:**
  o DO NOT attempt to open the accumulator without completely discharging the gas
  o Attach the proper charging and gauging unit.
  o Completely relieve the gas precharge.

6

6

Case Study for Accumulator Disassembling:

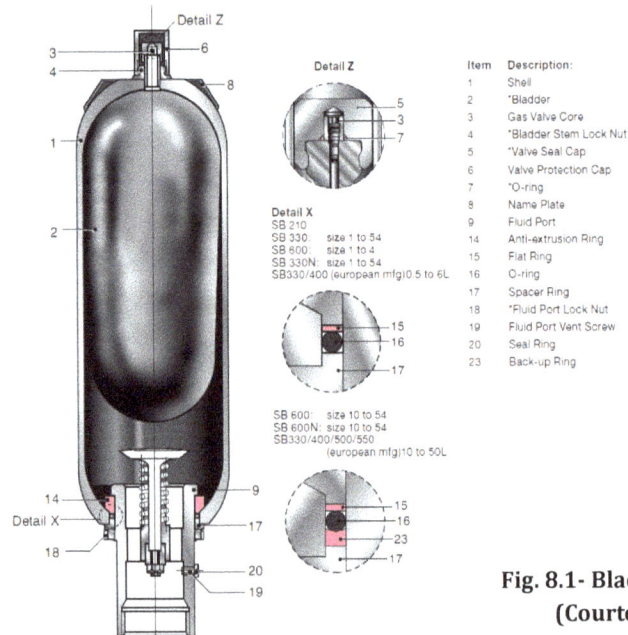

Item	Description:
1	Shell
2	*Bladder
3	Gas Valve Core
4	*Bladder Stem Lock Nut
5	*Valve Seal Cap
6	Valve Protection Cap
7	*O-ring
8	Name Plate
9	Fluid Port
14	Anti-extrusion Ring
15	Flat Ring
16	O-ring
17	Spacer Ring
18	*Fluid Port Lock Nut
19	Fluid Port Vent Screw
20	Seal Ring
23	Back-up Ring

Detail X
SB 210
SB 330: size 1 to 54
SB 600: size 1 to 4
SB 330N: size 1 to 54
SB330/400 (european mfg)0.5 to 6L

SB 600: size 10 to 54
SB 600N: size 10 to 54
SB330/400/500/550
(european mfg)10 to 50L

**Fig. 8.1- Bladder Accumulator
(Courtesy of Hydac)**

7

7

Example of a given instructions for disassembling a bladder accumulator (Shown in the previous figure):

A:

o Place the accumulator in a vice or secure it to a workbench.

o Remove **valve protection cap** (item 6) and

o Unscrew **valve seal cap** (item 5).

o Attach the proper Charging and Gauging Unit.

o Completely relieve the gas precharge

o Remove **gas valve core** (item 3) by using the gas valve core tool.

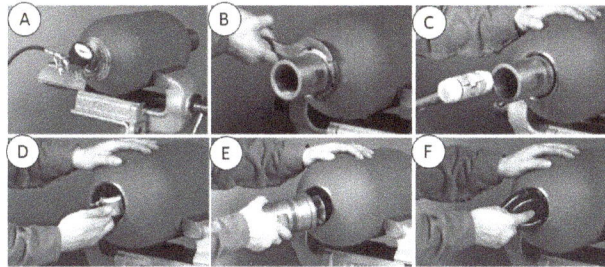

**Fig. 8.2- Bladder Accumulator Disassembling Instructions
(Courtesy of Hydac)**

8

8

B:

o Unscrew **vent screw** (item 19) and

o Remove **seal ring** (item 20).

o Unscrew **lock nut** (item 18) by using spanner wrench.

o Remove **spacer ring** (item 17).

o If necessary, tap spacer ring with a plastic hammer to loosen.

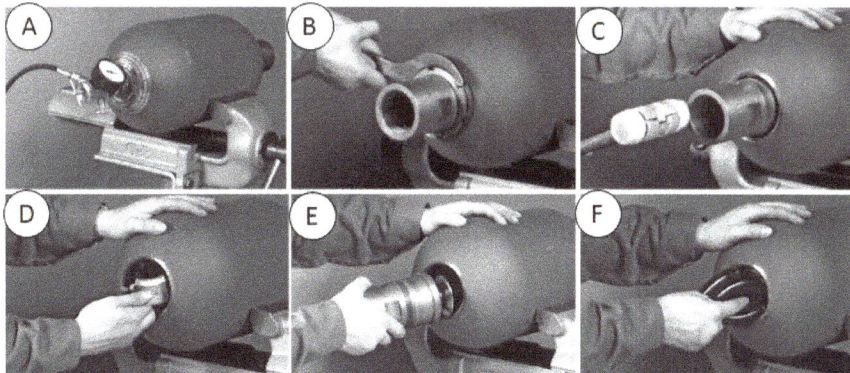

9

9

C:

- ○ Loosen **fluid port** (item 9) and push it into the shell.
- ○ Remove **back-up ring**, (item 23) where applicable,
- ○ Remove **O-ring** (item 16) and
- ○ Remove **flat ring** (item 15) from fluid port.

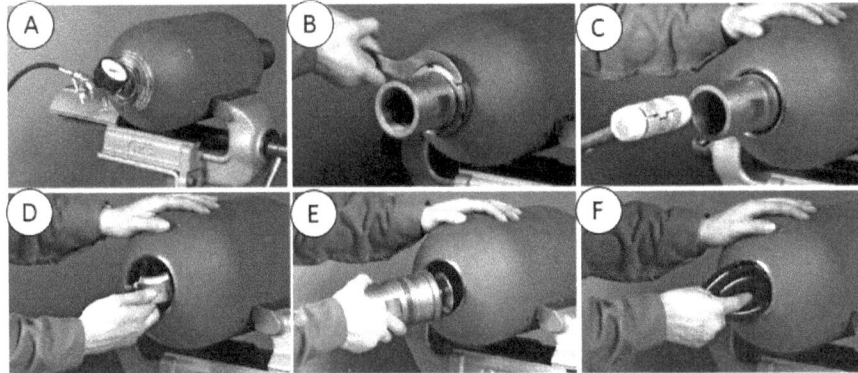

10

10

D: Pull **anti-extrusion ring** (item 14) off fluid port and remove it through fluid side opening by folding it in half.

E: Remove **fluid port** (item 9).

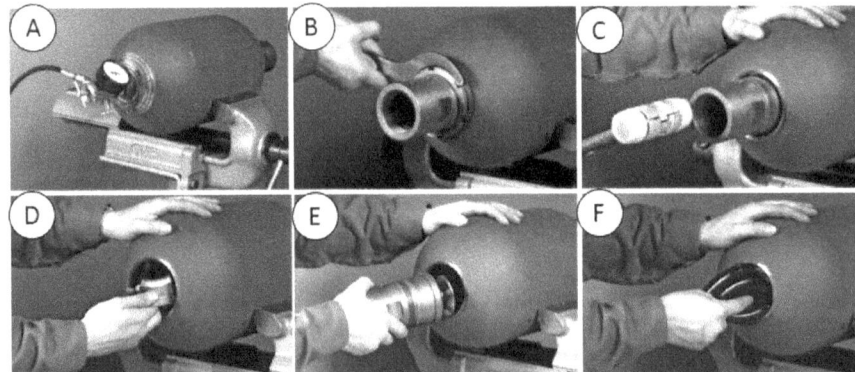

11

11

F:

- o Remove **bladder stem lock nut** (item 4) and **name plate** (item 8) from the gas side.
- o Remove **bladder** (item 2) from fluid side. It may be necessary to fold the bladder lengthwise to remove it.

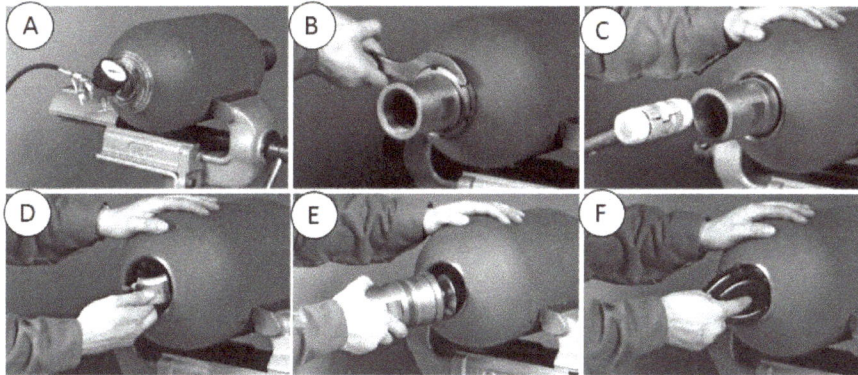

12

12

8.3.2- Inspection Instructions

- ▪ **Shell Inspection:**
  - o Interior (debris, rough spots, or damage marks).
  - o Exterior (any sign of damage such as scratches).
  - o If any interior or exterior sign of damage is found, contact the supplier for proper repair or replacement instructions.

- ▪ **Bladder:**
  - o Visual inspection (lateral grooves and deep chafe marks).
  - o If any are found → replace the bladder.

  - o Fill the bladder with nitrogen or compressed air to its natural shape
  - o Inspect for leakage.
  - o If leakage occurs, → first check the gas valve and replace it if necessary.
  - o If leakage still occurs, → then the bladder must be replaced.

  - o **Note:** Bladders can't be repaired or reused

13

13

- **Fluid Port:**
  - Visual inspection (poppet, threads, sealing surfaces, seats, and anti-extrusion ring) for any sort of damage.
  - If any damage is found, the fluid port should be replaced.

- **Seals:**
  - New seals should always be used whenever reassembling any bladder accumulator.

- **Other Parts:**
  - Visually inspect for damage and replace if necessary.

14

14

8.3.3- Assembly Instructions

Assembling: usually assembling is the reverse operation of disassembling.

Safety Block: An accumulator isn't a line mounted type of component.

- Use only fittings that are originally specified by manufacturer.
- Accumulators shall not be modified by machining, welding.

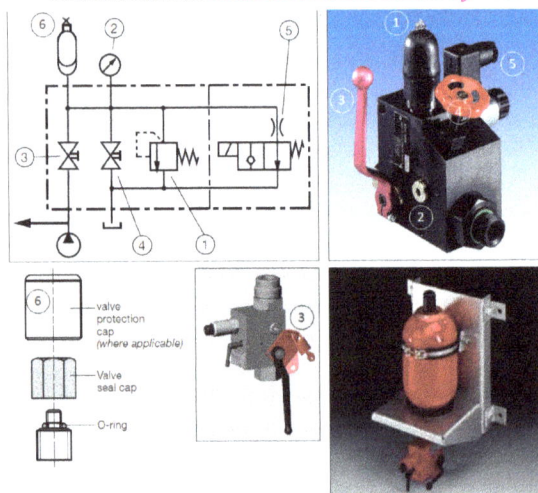

Fig. 8.3- Safety Base (Manifold) for an Accumulator (Courtesy of Hydac)

15

15

Mounting Guidelines:

- **Mounting Position:**
  - Best and safest position is vertical with gas valve upwards.
  - **Horizontal position:**
  - → Bladder friction with the internal surface.
  - → Trapped fluid → bladder distortion.
  - → Accelerated seal wear.
  - → **If** gas valve failed → accumulator acts as a rocket.

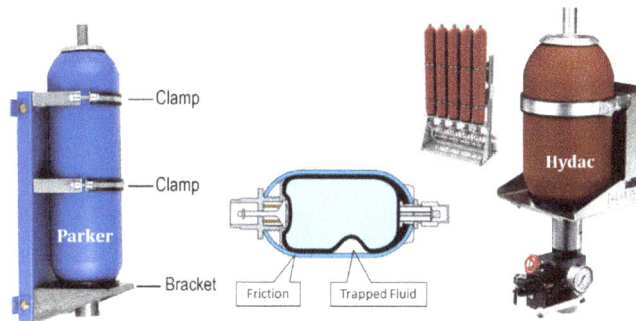

Fig. 8.4- Proper Mounting of Accumulators

16

16

- **Mounting Location:** Accumulators should be located in a place away from movement of other components with the least possibility of physical damage.

- **Space Above the Accumulator:** Maintain a space of at least 20 cm above the gas valve for testing and recharging services.

- **Supports:** Gas-loaded accumulators and any associated pressurized components shall be supported in accordance with the instructions of the accumulator supplier. However, accumulators must be well clamped. It is absolutely forbidden welding of supports or machining on the accumulator shell.

- **Heat Shield:** Accumulators should not be exposed to direct sunlight. Install a sunshade. Protect them with a baffle from other radiant heat sources such as die casting machines, furnaces, steam boilers, etc.

  V674-Accumulator Maintenance-Hydac-Bladder (7.5 min)

  V258-Accumulator Maintenance –Toubl (7.5 min)

17

17

8.3.4- Charging Instructions

- **Gas Type:**
  - Never use Oxygen or air.
  - Charge only with Nitrogen (N_2) with highest purity (class 4) of purity level 99.99% by volume.

- **Initial Gas Pressure:**
  - Unless otherwise stated by the manufacturer $\rightarrow$ the rule of thumb:
  - For Energy Storage Applications $P_0 = 0.9\ P_1$ **(Bladder Acc.)**
  - For Energy Storage Applications $P_0 = P_1 - 100$ **psi (Piston Acc.)**
  - For Shock Absorption Applications $P_0 = 0.9\ P_m.$
  - For Pulsation Damping Applications $P_0 = 0.9\ P_m.$
  - Where P_1 = Minimum system pressure.
  - and P_m = median system pressure = $(P_1 + P_2)/2$

18

18

- **Maximum Gas Pressure:**
  - Can't exceed the maximum system pressure.
  - Can be adjusted by a separate PRV mounted on the accumulator base.
  - Must respect the compression ratio allowed for the type of the acc.
  - Compression Ratio = $P_{max}/P_{min} = P_2/P_1$
  - Unless otherwise stated by the manufacturer $\rightarrow$ the rule of thumb:
    - ❖ Compression Ratio for Piston Accumulators, (8-9).
    - ❖ Compression Ratio for Bladder Accumulators (4-5).
    - ❖ Compression Ratio for Diaphragm Accumulators(1.5-2).

19

19

- **Temperature Effect:**

o Initial gas pressure should be adjusted for max temperature as follows:

o **Fahrenheit:** $P_{0,T0} = P_{0,T2} \times (T_0 + 460)/(T_2 + 460)$

o **Celsius:** $P_{0,T0} = P_{0,T2} \times (T_0 + 273)/(T_2 + 273)$

o Where:

o T_0 = Precharge temperature.

o T_2 = Maximum operating temperature.

o $P_{0,T0}$ = Gas precharge Pressure at precharge temperature.

o $P_{0,T2}$ = gas precharge pressure at maximum operating temperature.

20

20

- **Charging Device:**
- Charging and Gauging kits are available with male or female cap nut.

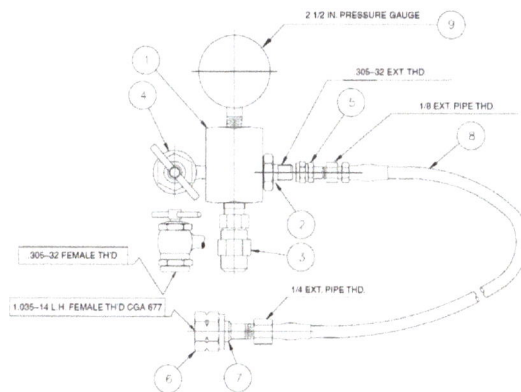

1	Adapter Cross
2	Gas Valve Assembly
3	Bleed Valve
4	Gas Chuck
5	Valve Connector
6	Nitrogen Bottle Nut with LH Connection
7	Gas Tank Nipple (CGA 677)
8	Charging Hose (10 ft.)
9	2.5" Pressure Gauge

**Fig. 8.5- Accumulator Charging Kit
(Courtesy of Parker)**

21

21

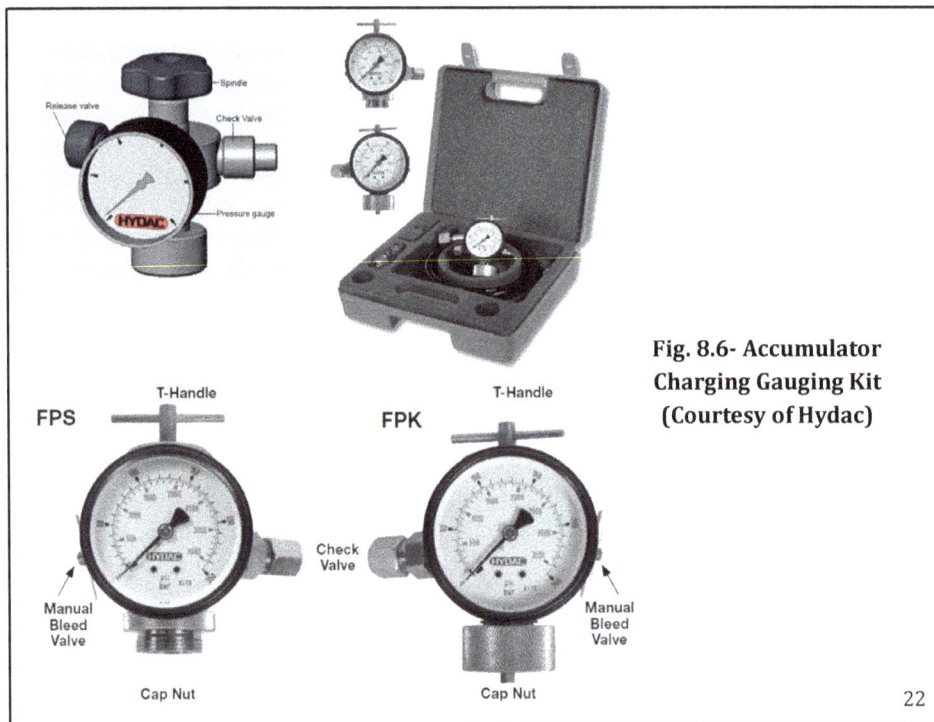

Fig. 8.6- Accumulator Charging Gauging Kit (Courtesy of Hydac)

22

• **Charging Process:**

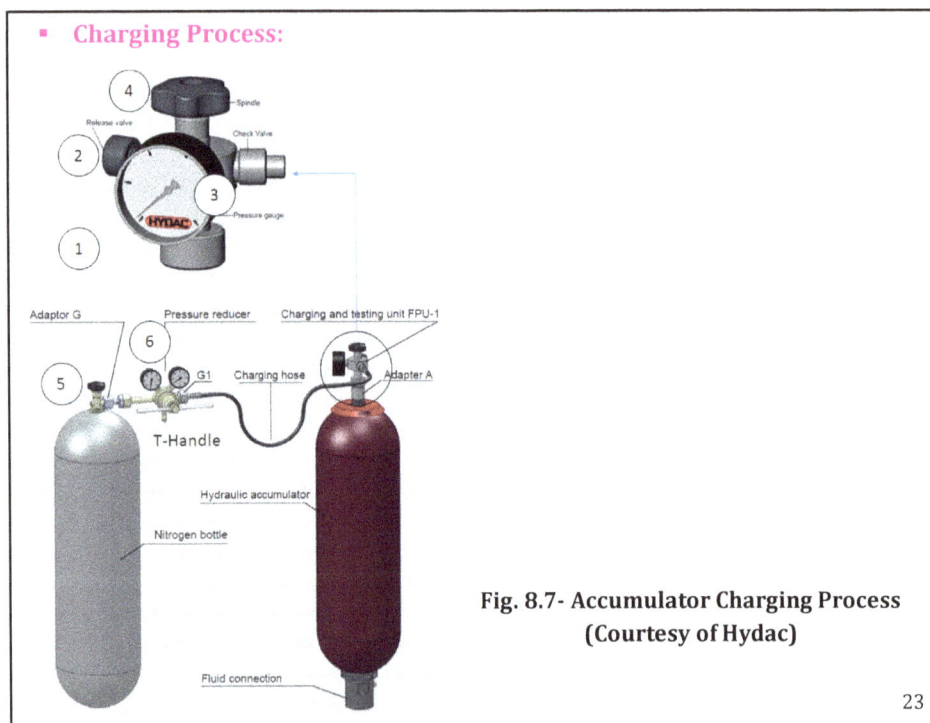

Fig. 8.7- Accumulator Charging Process (Courtesy of Hydac)

23

Connection to Accumulator:

- Make sure oil side of the accumulator is at zero pressure.
- Remove the sealing cap of the accumulator's gas valve.
- Make sure the T-Handle (6) on the charging unit is fully closed.
- Connect the charging kit (1) to the accumulator.
- Open the knop (4) → pressure gauge (3) →
- measure the current precharge pressure
- If > specified → press and release button (2) and lower the pressure.
- If < specified → proceed to the following step.

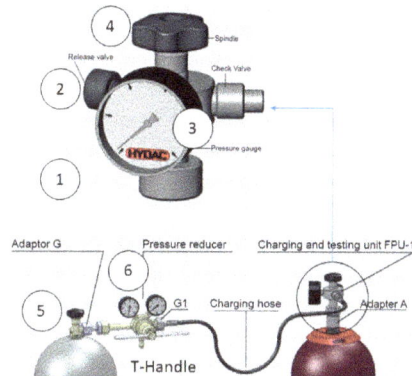

Connection to Nitrogen Cylinder:

- Make sure that the shut-off valve of the nitrogen bottle (5) is fully closed.
- Connect the other end of the charging assembly to the nitrogen bottle.

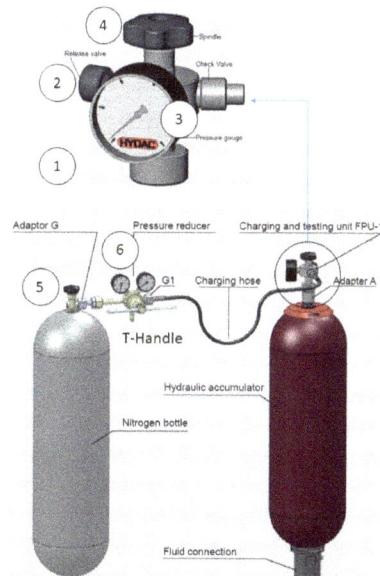

Charging the Accumulator:

o Gradually and slowly open the shut-off valve of the nitrogen bottle (5).

o Adjust the T Handle on the pressure regulator (6) to desired precharge value.

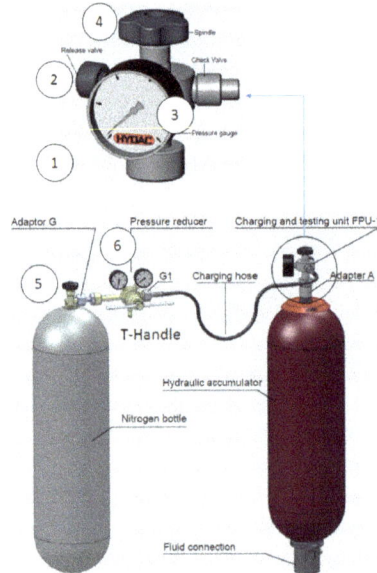

26

26

o Allow gas to flow slowly into the accumulator.

o The first 20-25 psi should take 2-3 minutes.

o Then adjust the charge rate to be one minutes for every increase of 25 psi.

o For every 25 psi, close the shutoff valve (5), give 5 seconds for the pressure to stabilize, and check the charge pressure at the pressure gauge.

o Repeat the previous two steps until the accumulator is charged to the assigned initial gas pressure.

o If needed, reduce excess charge pressure by carefully open the bleeding release valve (2) to get the precharge pressure reduced.

o Wait for proper time until the accumulator is cooled to the room temperature and the initial gas pressure to reach equilibrium:

➢ Adjusting an existing gas initial gas pressure → allow 5 to 10 minutes.

➢ Charging an empty accumulator → allow 20 to 30 minutes.

27

27

○ Close the shut-off valve (5) of the nitrogen cylinder and the knob (4).

○ Disconnect the charge assembly from the nitrogen bottle then from the accumulator.

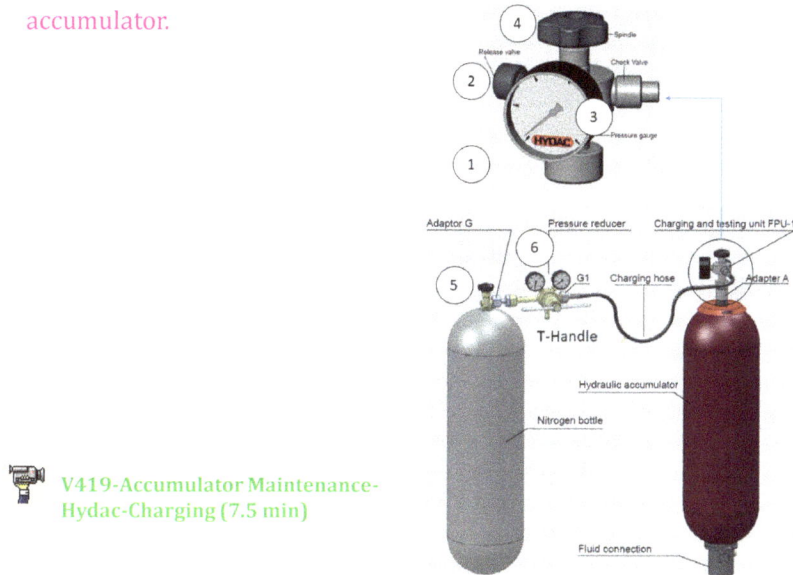

V419-Accumulator Maintenance-Hydac-Charging (7.5 min)

28

8.4- BP-Accumulators -04-Standard Tests and Calibration

- Accumulators are tested by the manufacturer.
- It is not the responsibility of end users to perform such tests.
- However, a test certificate should be provided by the manufacturer and must be kept in a safe place. The following set of bullets provide guidelines for testing an accumulator:

Testing Precharge Pressure:

Turn the system off → use a charging rig with a gauge.

29

Testing Working Temperature:

- Normal Pressure → fluid chamber (bottom half or two-thirds).
- Overcharged → heat is concentrated at the very bottom.
- Undercharged → heat is from top to bottom.
- No heat is measured → that may b because:
  o Automatic discharge device is stuck open.
  o The precharge pressure > pump compensator or PRV setting.

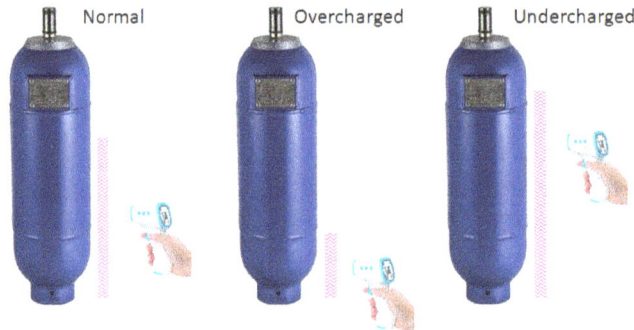

Fig. 8.8- Accumulator Temperature Test

30

30

8.5- BP-Accumulators -05-Transportation and Storage

- **Transportation:**
  o Accumulators should be gas and fluid discharged.
  o If charged, gas precharge should not exceeds 150 psi (10 bar) and proper labeling should be used to indicate the charging pressure.
  o Keep the isolation valve closed.
  o Review instructions for lifting. However, a gas valve or fluid port on an accumulator shouldn't be considered as liftings points in any case.

31

31

- **Storage:**
  o Accumulator must be discharged from both gas and oil before storage.
  o Piston accumulators are recommended to be stored in vertical position to prevent the seals from developing a set (flat spot) on the side that the piston weight is exerted.
  o Accumulators should be stored in a cool, dry place away from sun, ultraviolet and fluorescent lights as well as electrical equipment. Direct sunlight or fluorescent light can cause the seals to dry out.
  o The ideal temperature for storage is 70°F.

32

32

Chapter 8 Reviews

1. Which of the following information should be marked on the accumulator?
   A. Size of the pump in the system in which the accumulator is installed.
   B. Total shell volume (*Nominal Volume*) of the accumulator, expressed in liters or gallons.
   C. Setting of the pressure relief valve in which the accumulator is installed.
   D. Oil volume required t fill in the cylinder in connection with the accumulator.

2. Which of the following practices is highly prohibited when maintaining an accumulator?
   A. Completely discharge fluid before maintenance.
   B. Completely discharge gas before maintenance.
   C. Use proper charging and gauging unit when charging the accumulator.
   D. An accumulator is a simple device and no required experience or training is required for accumulator maintenance, just follow instructions.

3. Which of the following practices is highly recommended when assembling an accumulator?
   A. It doesn't matter whether the accumulator is assembled in a horizontal or a vertical position.
   B. An accumulator isn't a line-mounted hydraulic component. It should be assembled on a safety block specified by the accumulator manufacturer.
   C. To support an accumulator firmly to a machine body, the shell may be welded to a fixed frame of the machine.
   D. An accumulator isn't affected by the surrounding temperature because the thick wall of its shell. So, no shield from external heat sources during assembly,

4. A bladder accumulator is used in shock absorbing system. Initial gas pressure of such an accumulator should be?
   A. 80% of minimum system pressure.
   B. 90% of minimum system pressure.
   C. 90% of median system pressure.
   D. Minimum system pressure minus 100 psi.

5. Recommended compression ratio of a piston accumulator is?
   A. 1.5-2.
   B. 4-5.
   C. 8-9.
   D. >15.

Chapter 8 Assignment

Student Name: --- Student ID: ------------------

Date: -- Score: ------------------------

Question: List the information that should be marked on the accumulator.

Chapter 9
Maintenance of Reservoirs

Objectives:

This chapter provides guidelines for **reservoirs** selection, replacement, maintenance scheduling, installation, testing, storage and transportation. This chapter is supported by examples and figures granted by leading fluid power manufacturers.

Brief Contents:

9.1- BP-Reservoirs-01-Selection and Replacement

9.2- BP-Reservoirs-02-Maintenance Scheduling

9.3- BP-Reservoirs-03-Installation and Maintenance

0

0

The following topics are discussed in Chapter 2 in Volume 4 "Hydraulic Fluids Conditioning" of this series of textbooks:

- Contribution of Hydraulic Reservoirs
- Configurations of Hydraulic Reservoirs
- Construction of Hydraulic Reservoirs
- Design of Hydraulic Reservoirs
- Hydraulic Reservoir Design Case Study

The following topics are discussed in Chapter 9 in Volume 6 "Troubleshooting and Failure Analysis" of this series of textbooks:

- Hydraulic Reservoirs Inspection
- Hydraulic Reservoirs Troubleshooting
- Hydraulic Reservoirs Failure Analysis

1

1

9.1- BP-Reservoirs-01-Selection and Replacement

When selecting or replacing an existed reservoir →

- New reservoir should have the same (physical size, oil volume, air space, ground clearance, surface area, interior design) → to keep same reservoir contribution to the system.

- DO NOT change the placement of the suction and the return lines.

- DO NOT change the distribution of oil heaters inside the reservoir.

- Keep same plumping connections.

- Keep same electrical connections (fluid level, fluid temperature, etc.)

2

9.2- BP-Reservoirs-02-Maintenance Scheduling

Unless otherwise stated by components and systems manufacturer:

#	Preventive Maintenance Actions	Daily	Weekly	Monthly	Biannually	Annually
1	Check and repair any source of external leakage	✔	✔	✔	✔	✔
2	Check oil level in the tank and make up oil of needed	✔	✔	✔	✔	✔
3	Clean around and the outer surface including the sight glasses		✔	✔	✔	✔
4	Check proper connection to hydraulic lines.			✔	✔	✔
5	Check electrical connections			✔	✔	✔
6	Check the state of the breather filter and the presence of the filler cap (Note 1).			✔	✔	✔
7	Wash the suction strainers				✔	✔
8	Reservoir Major Maintenance (Note 2).: • Drain the oil • Clean inside the reservoir • Repaint with compatible paint • Change the cover gasket • Refill the reservoir • Clean/Replace air filter • Clean the strainer				✔	✔

Table 9.1- BP-Reservoirs-02-Maintenance Scheduling

3

Note 1:

o   Breathers are consumable elements must be changed periodically.

o   If no information about when to replace → rule of thumb:

   ➤ Every three months in dirty environments (e.g. foundries).

   ➤ Every six months in cleaner environments.

Note 2:

o   The best time to check a reservoir is during the routine shutdown time of the equipment.

4

4

9.3- BP-Reservoirs-03-Installation and Maintenance

Well-designed reservoirs can cause problems if not installed properly.

Clean Outside the Reservoir:

o   Dirt accumulated on outside surface → Heat dissipation ↓

o   → fluid degradation ↑ → sludge and varnish formation ↑.

o   Cleaning outside the reservoir must include cleaning all other devices that are assembled on the reservoir such as gauges, level indicators, etc.

Clean Inside the Reservoir: Reservoir cleanliness is a vital maintenance task because:

o   Any contaminants left in the reservoir will circulate in the system.

o   pumps and other components are designed with close clearances to operate in the 2000 to 5000 psi range, and they have become much more sensitive to dirt.

5

5

- **Drainage:** A reservoir must be completely drained.
- **Sandblasting:** Properly wiped the left-over dust.
- **Flushing:** Flush with oil or a solvent to remove all traces of sand. Review flushing process in Volume 3.
- **Painting:**
  o Painting steel to prevent rust.
  o Painting inside and outside surfaces.
  o Paint must be compatible with the fluid being used and the storage and operating system temperature.

Before Cleaning   After Cleaning

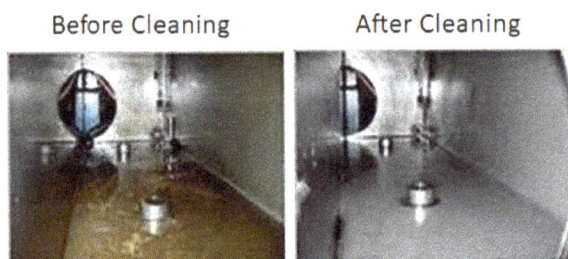

Fig. 9.1- Cleaning of Hydraulic Reservoirs

6

6

Clean Suction Strainers:
o Outer surface is coated by brown varnish→ oil is overheated.
o Clean washable strainers using approved compatible solvent.
o Dry a strainer by blowing clean air from inside to outside.

Cleaning Solvents:
o Make sure it's recommended for hydraulic systems and specified by the system manufacturer.
o Even very small amounts of the wrong solvent can affect certain hydraulic fluid additives.

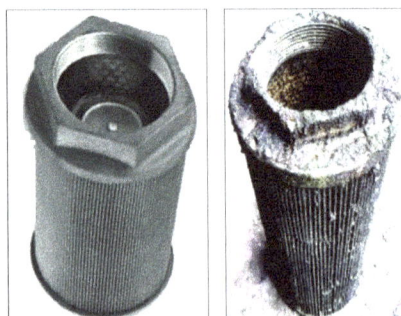

Fig. 9.2- Clean versus Dirty Suction Strainer

7

7

Cleaning Towels:
o   Make sure to use an industry specified and lint-free towels.

Cleaning/Replace Air Breather Cartridge:
o   The visual indicator shows the blockage level of the air breather element.

Change Gaskets:
o   Upper cover or the side cover (manhole) or flanges are removed →
o   gasket must be replaced.
o   Seals and gaskets must also be compatible with fluid.

Clean Around the Reservoir:
o   Oil around the reservoir→ :

➢   Check for leaks around fittings and make the necessary repairs.

➢   Clean up leakage immediately to avoid slippage or fire potential.

➢   Do not return any spillage to the reservoir.

➢   Put mopped spillage and associated materials into proper disposal containers.

8

8

Space Around:
o   Reservoirs must be placed in a well vented space to allow heat dissipation.

o   DO NOT locate the reservoir under benches, tables, or in cabinets.

o   If needed, use an industrial fan to force air around the reservoir.

Space Below:
o   Ground clearance at least 15 cm (6 inches).

Shield from Heat:
o   Shield reservoirs from heat sources (furnaces, steam pipes, steel mills, etc.).

o   Shield reservoirs from harsh environment (direct sun light, snow, very cold or very hot weather, rain, flood, dust storm, etc.).

o   Connecting pipes should also be shielded.

9

9

Refill or Making up the Reservoir:

o   Thinking a new fluid is clean, is a common mistake.

o   Never add new fluid or make up exiting fluid without passing it through a filter.

o   Unless otherwise is stated, use filter rating of 10-micron & Beta Ratio 75.

o   NEVER FILL TO THE TOP! Oil expanding under working temperatures will overflow the reservoir.

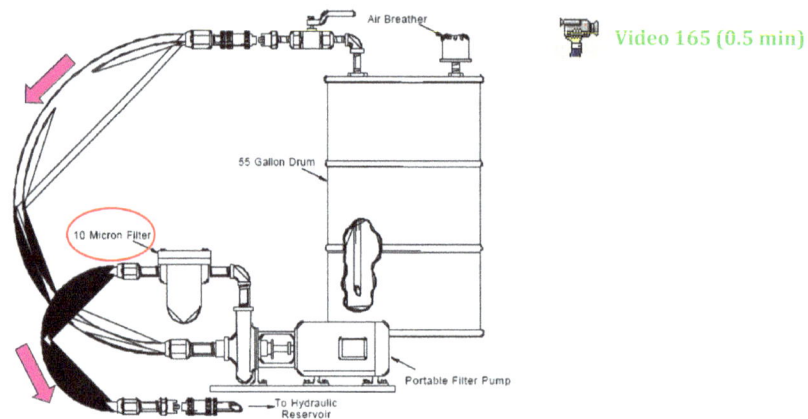

Video 165 (0.5 min)

Fig. 9.3- Best Practices for Filling Hydraulic Reservoirs   10

10

Chapter 9 Reviews

1. Which of the following reservoir design parameters affects the reservoir contribution in heat dissipation?
   A. Oil volume in reservoir.
   B. Ground space.
   C. Volume of oil.
   D. All of the above.

2. Which of the following maintenance action supposed to be performed more frequent than any other actions?
   A. Checking for oil level and external leakage.
   B. Check proper connections with hydraulic lines.
   C. Check proper electrical connection.
   D. Wash suction strainer.

3. If a suction strainer is coated by a layer of varnish, this means that?
   A. Dirt is accumulated on reservoir surface.
   B. Reservoir isn't able to dissipate heat.
   C. Hydraulic fluid is over heated.
   D. All of the above.

4. Which of the following statement is considered FALSE?
   A. New fluid in a drum is clean and may be used directly to make up fluid in a reservoir.
   B. Even new fluid must pass through a proper filter to make up fluid in a reservoir.
   C. Never fill a reservoir to top, a proper air space must be maintained.
   D. All of the above is False.

5. Ground clearance under the reservoir should be at leas?
   A. 5 cm.
   B. 10 com.
   C. 15 cm.
   D. 20 cm.

Chapter 9 Assignment

Student Name: --- Student ID: ------------------

Date: -- Score: -----------------------

Question: Explain best practices for refilling or making up hydraulic fluid in a reservoir.

Answer:

Chapter 10
Maintenance of Transmission Lines

Objectives:

This chapter provides guidelines for **transmission lines** selection, replacement, maintenance scheduling, installation, testing, storage and transportation. This chapter is supported by examples and figures granted by leading fluid power manufacturers.

Brief Contents:

10.1-BP-Transmission Lines-01-Selection and Replacement

10.2-BP-Transmission Lines-02-Maintenance Scheduling

10.3-BP-Transmission Lines-03-Installation and Maintenance

10.4-BP-Transmission Lines-04-Standard Tests and Calibration

10.5-BP-Transmission Lines-05-Transportation and Storage

0

0

The following topics are discussed in Chapter 3 in Volume 4 "Hydraulic Fluids Conditioning" of this series of textbooks:
- Basic Types and Contribution of Hydraulic Transmission Lines
- Sizing of Hydraulic Transmission Lines
- Rated Pressures for Hydraulic Lines
- Hydraulic Pipes
- Hydraulic Tubes
- 3.6- Hydraulic Hoses
- Flanges for Transmission Line Connections
- Rubber Expansion Fittings
- Test Points
- Pressure Measurement Hoses
- Manifolds

The following topics are discussed in Chapter 10 in Volume 6 "Troubleshooting and Failure Analysis" of this series of textbooks:
- Hydraulic Transmission Lines Inspection
- Hydraulic Transmission Lines Troubleshooting
- Hydraulic Transmission Lines Failure Analysis

1

1

10.1-BP-Transmission Lines-01-Selection and Replacement

Transmission lines are originally specified to:

- Consider leak-free connections
- Laminar flow conditions.
- Comply with the operating conditions (pressure, temperature, fluid compatibility, etc.)

When selecting or replacing an existed line →

- **Size (based on inner diameter)** → laminar flow.
- **Temperature:** recommended range + thermoplastic type.
- **Application:** suction versus pressure hoses.
- **Pressure:**
  - look like each other DOES NOT mean they can replace each other.
  - Review maximum allowable pressure (static and dynamic).
- **Media (Fluid Compatibility).**
- **Electrically Nonconductivity.**
- **Length and type of end joints/fittings.**
- **Pipes and Tubes Material.**

2

2

10.2-BP-Transmission Lines-02-Maintenance Scheduling

Unless otherwise is stated by components and systems manufacturer:

#	Preventive Maintenance Actions	Daily	Weekly	Monthly	Biannually	Annually
1	Line visual inspection **(Note 1)**	✔	✔	✔	✔	✔
2	Inspect for leakage **(Note 2)**		✔	✔	✔	✔
3	Clean around and the outer surface including the end joints **(Note 3)**		✔	✔	✔	✔

Table 10.1- BP-Transmission Lines-02-Maintenance Scheduling

3

Note 1 (Visual Inspection):

Video 150 (1 min)

- Like a vehicle tire, a transmission line should be replaced based on given service life no matter how it looks like.

- A hose service life is generally shorter than hard tubing.

- Immediately shut down and replace the Hose (no matter what its lifetime is) if:
  - Kinked, crushed, flattened or twisted Hose.
  - Cracks from minimum bend radius exceeded and around fittings.
  - Brittleness or loss of flexibility.
  - Frayed protective layers or broken reinforcement layers.
  - Outer cover pulled back from the end of the coupling.
  - Fitting slippage on Hose.
  - Rusted, broken, cracked, damaged, leaking, or badly corroded Fittings.

4

4

Note 2 (Inspection for Leakage):
- Never tighten fittings while the system is under pressure.
- DO NOT use your hand. Avoid possible oil injection.
- Wear proper industrial gloves that prevent oil injection.
- Use proper technique (fluorescent dye and UV light).

Video 163 (1 min)

Video 658 (2.5 min)

Even a pinhole leak can release hydraulic fluid at a pressure high enough to penetrate both clothing and skin.

Never use your hand to check for hydraulic leaks. The results could cost you a limb or even your life.

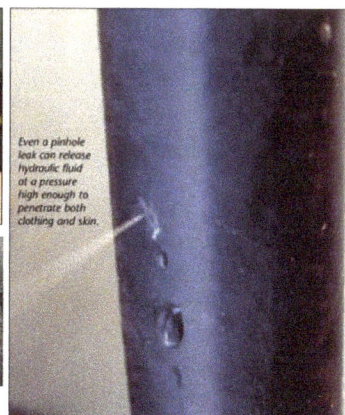

Fig. 10.1- Bet Practices for Hydraulic Transmission Line Leakage Inspection (www.bondfluidaire.com)

5

5

- If proper equipment isn't available, use a piece of wood or cardboard that is 1-2 feet long. Hold one end of the wood or cardboard, place the other end approximately 1-2 inches away from the inspected part of the line, and move it around the line. Video 149 (1 min)

- For challenges and best practices if fluid injection occurs, refer to section 1.14 in Chapter 1 (Hydraulic System Safety) of this textbook.

Note 3 (cleaning):
- Cleaning (outer surfaces + end joints) →
- better heat dissipation +
- maintain overall acceptable clean environment.

Fig. 10.2- Keep Outer Surfaces including End Joints (mac-hyd.com)

6

6

10.3-BP-Transmission Lines-03-Installation and Maintenance

- Proper installation and maintenance of transmission lines →
- trouble-free operation of the system.

BP-Transmission Lines-03-Installation and Maintenance:

1. Proper Line Cleaning before Assembling.

2. Proper Hose Crimping.

3. Proper Hose Routing.

4. Proper Hose Assembling.

7

7

10.3.1- Proper Line Cleaning before Assembly

- Contamination settled inside new and used hydraulic lines.
- One or combination of (Pickling, Flushing, and Projectile Cleaning).
- Detailed information in Volume 3 (Chapter 10).

Pickling (Cleaning by Acids):
- This process isn't applicable for flexible hoses.
- Producing hard lines involve harsh steps (e.g. extrusion, trimming, etc.).
- → built-in scales, greases, and many other contaminants left inside.
- Pickling liquids must be handled in accordance with OSHA requirements.
- Pickling process must be carried out by a special contractor.
- Pickling process consists of the following sequences:
  1. Degreasing all parts.
  2. De-rusting using commercial de-rusting solution.
  3. Rinsing in cold running water.
  4. Neutralizing in another tank using neutralizing solution.
  5. Rinsing in hot water.
  6. Drying by blowing dry, hot, filtered air.

8

8

Flushing (Cleaning by Hydraulic Fluids):
- This process (similar to kidney wash) is applicable for all types of lines.
- In first use, or after major maintenance, or after certain hours.
- Contaminants settled in lines:
  o During production, handling, and storage.
  o During operation because of low fluid velocity.
  o Lack of sufficient filtration.
- Flushing is a process where low viscosity fluid is circulated at high velocity and high temperature through the system to scavenge all contaminates.

9

9

Projectile Cleaning (Cleaning by Compressed Air):

- Projectile cleaning is a method of final cleaning of all types of lines.

- Commonly after hose cutting, tube flaring, or pipe threading.

- The easiness of the process makes it usable in field.

- The method is simply pushing a projectile inside the line using compressed air.

- Proper selection of quality of air, projectile size, and projectile type are essential for successful projectile cleaning process.

Video 138 (1.5 min)

10

10

10.3.2- Proper Hose Crimping

1. Setup the crimping machine referring to the manufacturer' instructions identifying the crimp specifications shown in the figure.
2. Select the proper die series. The dies are color coded and stamped on the top.
3. Before loading the die, brush the inside surface by a lubricant.
4. Load the selected dies into the crimper.
5. Place the die ring above the die.
6. Determine the hose insertion depth.
7. Insert the hose into the crimpable fitting until the insertion mark aligned with the end of the fitting.
8. Insert the hose into the die and properly align the coupling with the die fingers.
9. Finish the crimping, remove the hose, check the crimping diameter using a caliper.

11

11

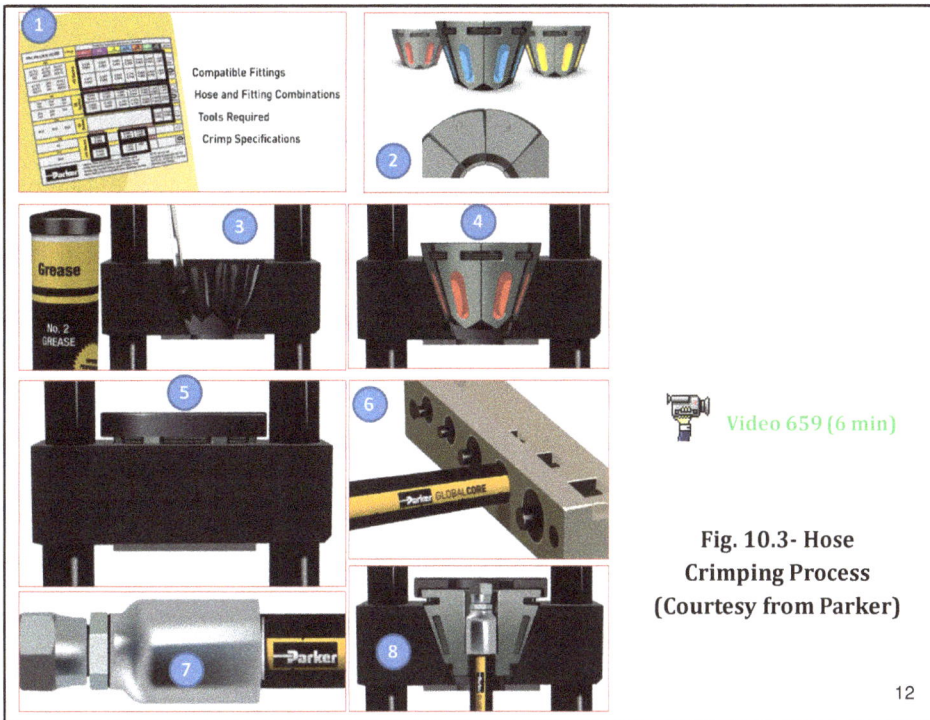

Video 659 (6 min)

Fig. 10.3- Hose Crimping Process (Courtesy from Parker)

Hose Insertion Depth:

- For hose Overall Assembly Length Cut Length → review Volume 4.
- Hose Insertion Depth must be checked and marked before crimping.
- Estimating the depth of the hose by eyes → fitting can blow off.

Fig. 10.4- Hose Crimping Machines (Courtesy from Gates)

1- Insert the hose into appropriate size die

2- Mark end of the hose to indicate Proper fitting insertion depth

3- Hose is ready for assembling

13

10.3.3- Proper Hose Routing

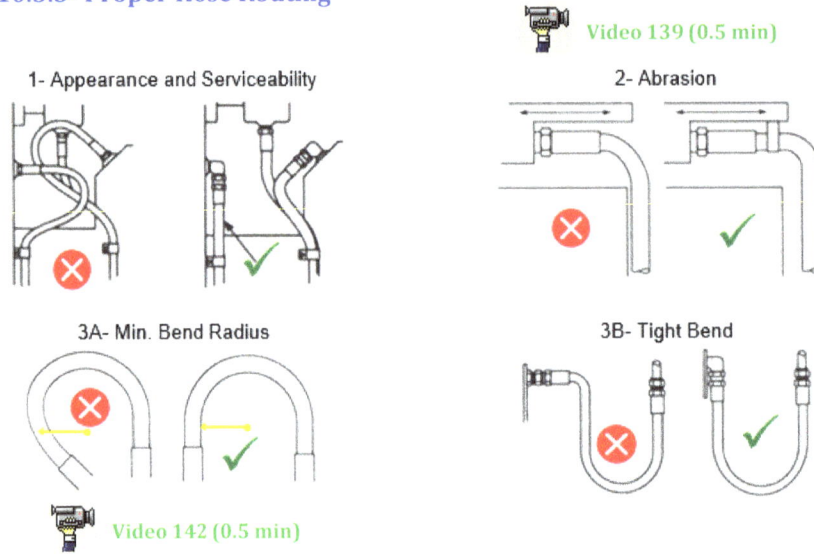

Fig. 10.5- Best Practices for Hydraulic Hose Routing

14

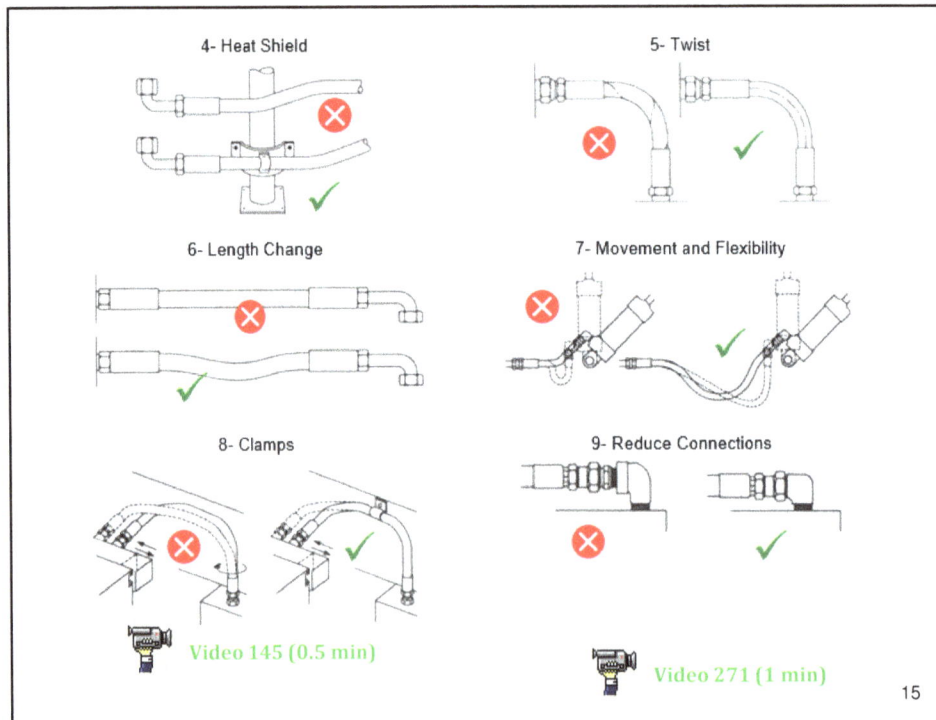

15

Separate Hydraulic Lines from Electric Lines:
- Intersection or contact with electrical lines → possible explosion or ignition in case of developing spark and oil spray.

Shield Hoses from Heat Sources: Used guards and shields to keep Hoses away from extreme temperatures or heat sources. Video 144 (1 min)

Protect the Hoses from Hazardous Material and Conditions: Use armures, guards, or sleeves to protect the hoses as needed. Also, keep hoses away from chemicals where possible

Quick-Disconnect Couplings: When assembling quick connect couplings, make sure they are matched. Ensure they are connected/disconnected properly. If in doubt, disconnect and re-connect the couplings again.

16

16

10.3.4- Proper Hose Assembling

1. Clean the surrounding area where connections are to be made.
2. Install adapters into ports (if used). Torque to specifications.
3. Verify length and correct routing.

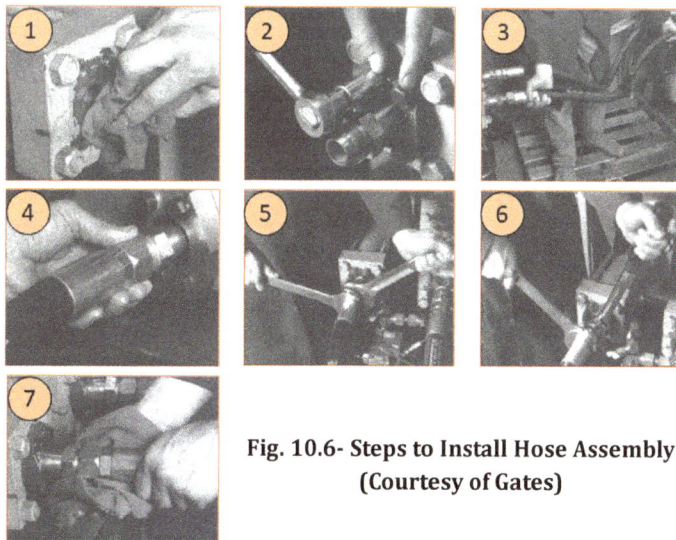

Fig. 10.6- Steps to Install Hose Assembly (Courtesy of Gates)

17

17

4. Thread one end of hose assembly (angled fitting first if found).
5. Thread other end of the assembly without twisting the hose.
6. Properly torque both ends with manufacturer's torque considered.
7. Run the hydraulic system to circulate oil and check for leakage.

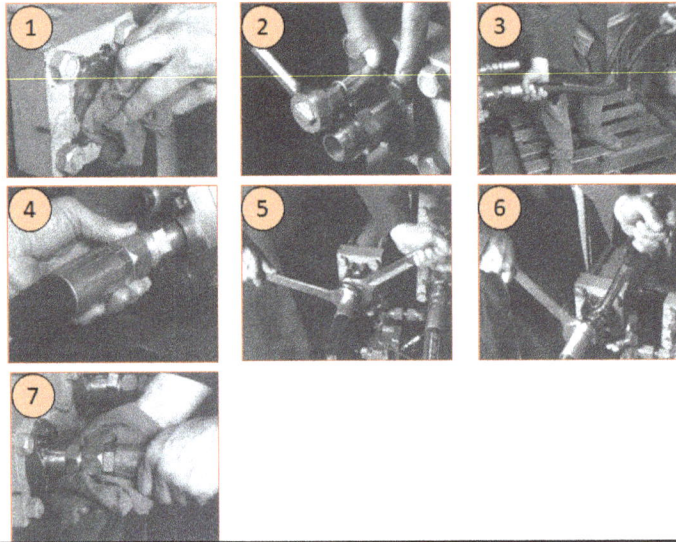

18

18

10.3.5- Assemble for Leakage Prevention

Fluid stains or puddles under hydraulic equipment or transmission lines → presence of a leak in the line.

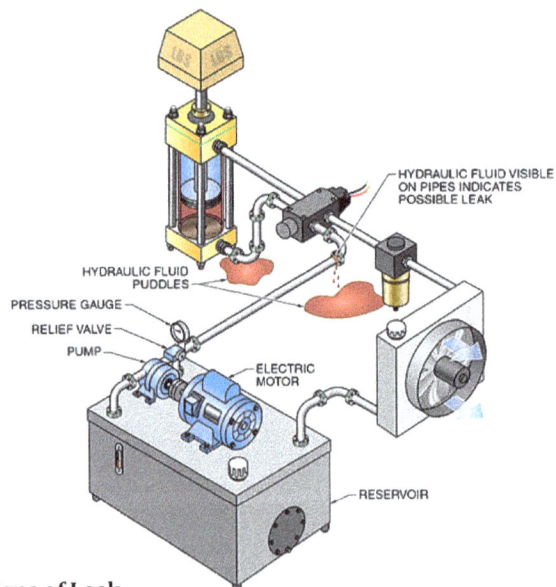

Fig. 10.7- Transmission Line Signs of Leak (Courtesy of American Technical Publisher)

19

19

Best Practices to Prevent Line Leakage:

Video 148 (2 min)

❖ **Quality of Product:**

▪ Never use galvanized steel or commercial "from-off-the-shelf" fittings.

▪ Fittings and crimpers are designed to work together as a system.

▪ Mixing of fittings and hoses form different suppliers → hose assembly leakage, hose separation or other failures.

❖ **Determining the Thread Type:**
▪ Threads of various fittings look similar.
▪ Mixing fittings of different thread types → leakage.

▪ **Thread Gauge:**
  o Number of threads per inch.
  o Holding the gauge in front of a lighted background → accurate measurement.

▪ **Caliper Measure:**
▪ Measure the thread OD and ID diameters.

20

Thread Gauge

Inside Thread Diameter

Outside Thread Diameter

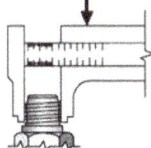

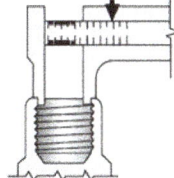

Vernier Calipers

Fig. 10.8- Determining the Thread Type of Hydraulic Fittings

21

❖ **Fittings Assembly Torque:**
- Over torqueing a fitting → it DOES NOT mean it seals better.
- Overtightening → overstressing or cracking.
- Never tighten a fitting while the pump is running.
- Review manufacturers specified assembly torque.
- The minimum value will create a leakproof seal under most conditions.

Size		Steel			
		Ft. Lbs.		Newton-Meters	
Dash	Inches	Min	Max	Min	Max
-4	1/4	10	11	13	15
-5	5/16	13	15	18	20
-6	3/8	17	19	23	26
-8	1/2	34	38	47	52
-10	5/8	50	56	69	76
-12	3/4	70	78	96	106
-16	1	94	104	127	141
-20	1-1/4	124	138	169	188
-24	1-1/2	156	173	212	235
-32	2	219	243	296	329

Table 10.2- Recommended Tightening Torque for 37° & 45°
(Machined or Flared Fittings) (www.new-line.com)

22

22

Size		Newton-Meters	
Dash	Inches	Min	Max
-4	1/4	14	16
-6	3/8	24	27
-8	1/2	43	54
-10	5/8	60	75
-12	3/4	90	110
-14	7/8	90	110
-16	1	125	240
-20	1-1/4	170	190
-24	1-1/2	200	245

Table 10.3- Recommended Tightening Torque for Flat-Face O-Ring Seal (Steel)
(www.new-line.com)

23

23

Size		Ft.Lbs. Working Pressures 4,000 Psi (27.5 Mpa) And Below		Newton-Meters Working Pressures 4,000 Psi (27.5 Mpa) And Below		Ft.Lbs. Working Pressures Above 4,000 Psi (27.5 Mpa)		Newton-Meters Working Pressures Above 4,000 Psi (27.5 Mpa)	
Dash	Inches	Min	Max	Min	Max	Min	Max	Min	Max
-3	3/16	–	–	–	–	8	10	11	13
-4	1/4	14	16	20	22	14	16	20	22
-5	5/16	–	–	–	–	18	20	24	27
-6	3/8	24	26	33	35	24	26	33	35
-8	1/2	37	44	50	60	50	60	68	78
-10	5/8	50	60	68	81	72	80	98	110
-12	3/4	75	83	101-1/2	113	125	135	170	183
-14	7/8	–	–	–	–	160	180	215	245
-16	1	111	125	150	170	200	220	270	300
-20	1-1/4	133	152	180	206	210	280	285	380
-24	1-1/2	156	184	212	250	270	360	370	490

**Table 10.4- Recommended Tightening Torque for SAE Steel O-Ring Boss
(www.new-line.com)**

24

24

❖ **Proper Assembling of Flareless Fitting:**

1. Properly burr the tube end, clean the burring products, and visually inspect the tube end to make sure it is burred properly.

2. Check the 90⁰ tube end trimming using proper tool.

3. Apply lubricant to the fitting.

4. Assemble the nut and the cutting ring with the tube.

5. Assemble the tube assembly with the fitting.

6. Tight by hand.

7. Tight by a wrench to torque specified by manufacturer.

8. Disassemble the tube assembly and check the proper attachment between the tube and cutting ring. The cutting ring may rotate in place around the tube but must not move axially along the tube. A special light source tool is used to make sure light isn't seen from between the tube and the cutting ring. If so, fluid leak will not occur.

25

25

Video 664 (3.5 min) Video 660 (3.5 min)

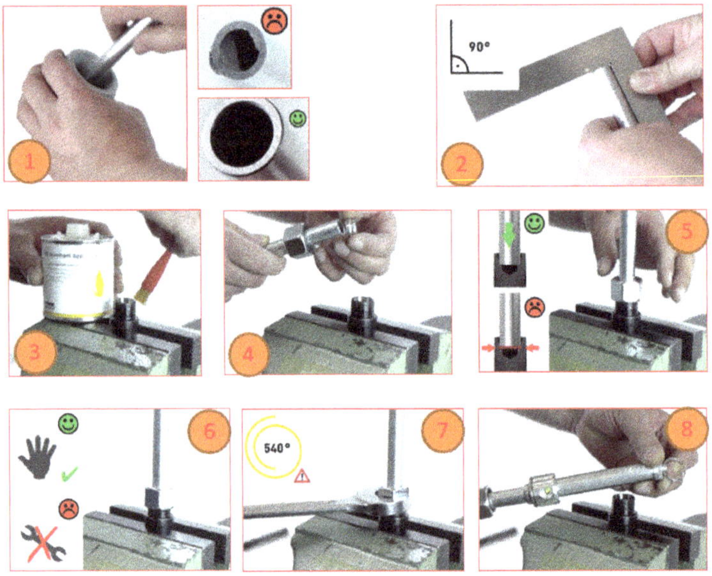

Fig. 10.9- Best Practices for Assembling Flareless Fitting (Courtesy from Parker)

26

26

❖ **Proper Tube Flaring:**

Video 240 (1 min)

- Improper tube flaring → cracked tube nose and fluid leakage.

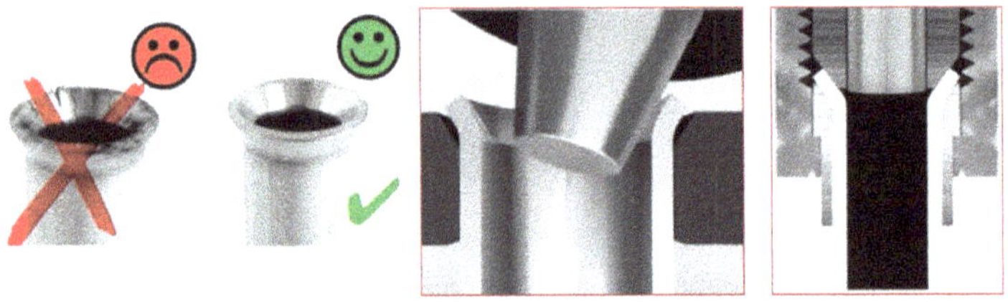

Video 655 (1 min)

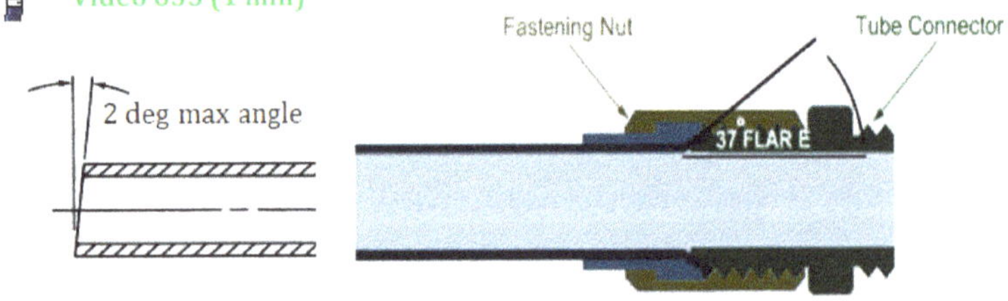

Fig. 10.10- Best Practices for Tube Flaring (Courtesy from Parker)

27

27

- Guidelines for proper tube flaring:

1. Select proper tubing (fluid compatibility, operating pressure, etc.)
2. Review (Volume 4) the standard flaring dimensions (single or double flare).
3. Cut the tube to required length.
4. End should be trimmed square within +/- 2° tolerance.
5. Properly burr O.D. and l.D. of tube.
6. Clean tube to remove all dirt from both O.D. and I.D. of tube.
7. Assemble tube nut and sleeve on tube.
8. Flare the tube end using the correct flaring tool (size & flare angle)
9. Inspect flare to the standard dimensions indicated in the tables.
10. Check flare for concentricity, thin out, cracks, nicks, loose slivers, burrs, pits or other defects which may prevent sealing.

28

28

❖ **Proper Assembling of Flared Fitting:**
- Manufacturer Specification (after tightening by hand):
  o Tightening torque OR how many faces the nut should rotate.

1- Clap the body on a Vice

2- Tightened by Hand

3- Tightened by Wrench to Specified Torque

Video 654 (2.5 min)

Fig. 10.11- Best Practices for Assembling 37° Flared Fitting (Courtesy from Parker)

29

29

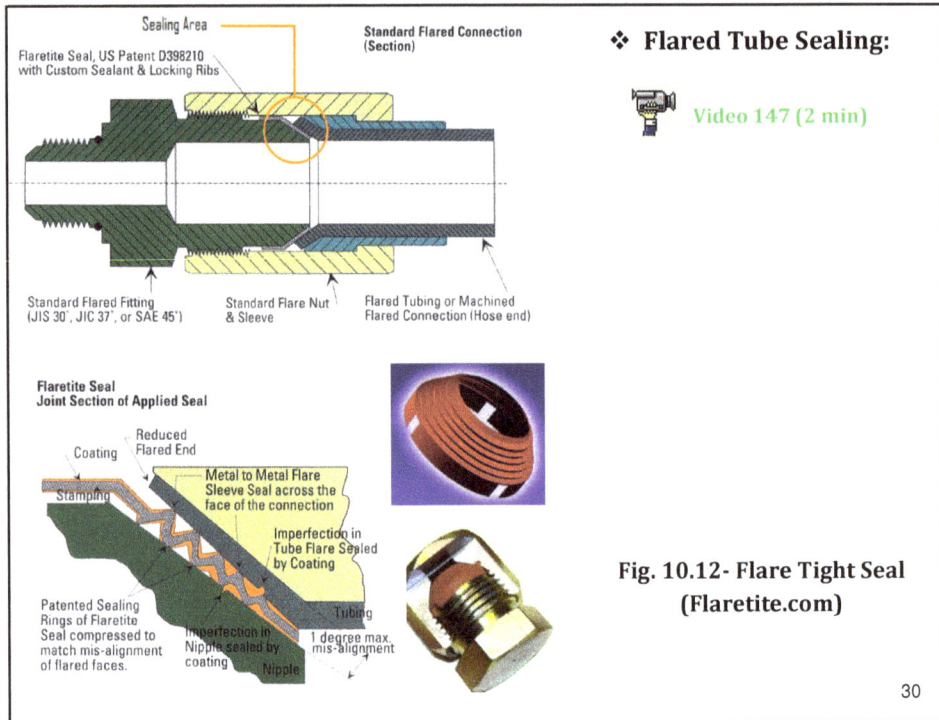

❖ **Flared Tube Sealing:**

Video 147 (2 min)

Fig. 10.12- Flare Tight Seal
(Flaretite.com)

30

❖ **Proper Tube Bending:**
- Minimum bend radius = f (tube size, pressure ratings, etc.)
- Minimum bend radius specified by manufacturer.

Video 657 (6 min)

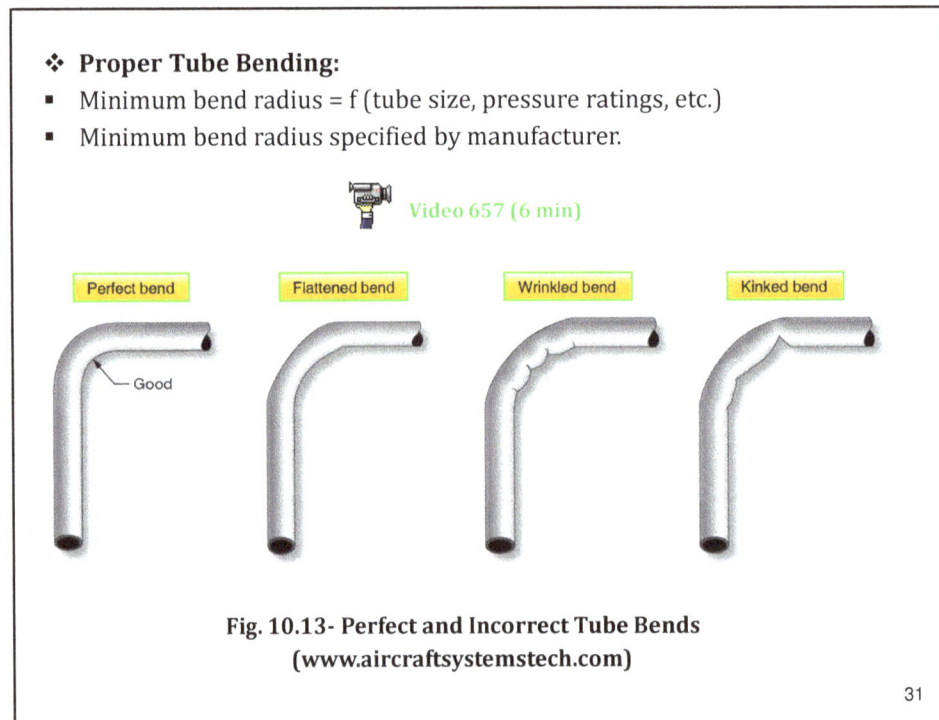

Fig. 10.13- Perfect and Incorrect Tube Bends
(www.aircraftsystemstech.com)

31

❖ **Proper Pipe Welding:**

▪ Surrounding components must be properly isolated.

▪ Welding personnel must be certified and well trained.

▪ Inspect after welding, sealing and internal surfaces shall be free of any visible detrimental foreign matter such as scale, burrs, swarf, etc.

❖ **Proper Pipe Assembly:**

❖ Leave the end pipe threads free from sealant or Teflon® tape to ensure that it does not contaminate the system.

Video 176 (1 min)

THREAD SEALANT		PIPE WRENCH
Apply thread sealant	Start fitting by hand	Tighten by pipe wrench

Fig. 10.14- Pipe Assembly (Courtesy from American Technical Publishers)

32

32

❖ **Transmission Line Proper Clamping:**

▪ **Unclamped Hydraulic lines →**

o Noise and vibration.

o If a pressurized line assembly blows apart → fittings are thrown off at high speed → risks of injuries.

o If one end of pressurized hose is accidentally detached → it whips like a snake with great force.

Video 650 (4 min)

33

33

- **Best practices for hose clamping:**
  - Do Not temporarily drop a return line hose into the reservoir.
  - Do Not allow hoses to drag in the dirt or lay on the ground.
  - Clamp hoses to the structure; not to components to minimize vibration.
  - Clamp far from moving parts.
  - Clamp to avoid surface abrasion (Fig. 10.15).
  - Use hose restraints (Fig. 10.16) to prevent injury when a hose is blown.

Fig. 10.15- Proper Hose Clamping

Fig. 10.16- Restrains for Hoses Clamping

34

34

- **Best practices for tubes/pipes clamping:**
  - Different standard tube clamping collars are available.
  - Material (polypropylene, aluminum or steel with plastic supports).
  - They are fastened via a weld plate, a thread, or a guide.

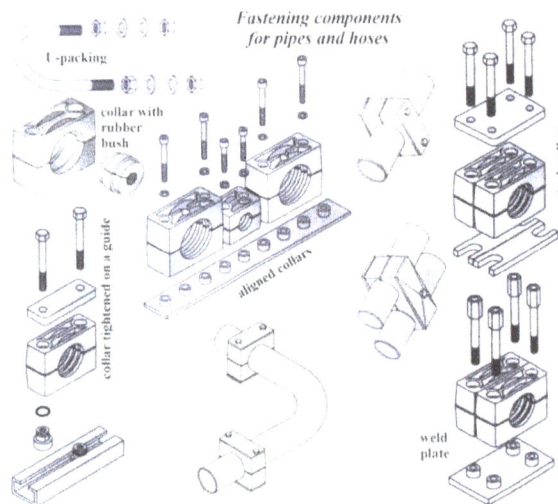

**Fig. 10.17- Clamping Collars
for Tubes and Pipes
(Courtesy of Assofluid)**

35

35

- Random clamping distance → generating of noise and vibration.

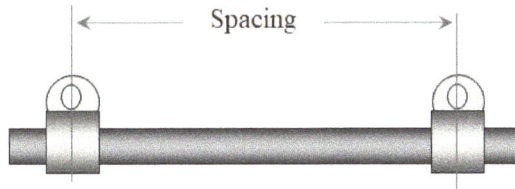

Spacing

Table-11-1			
TUBE O.D."	EQUIVALENT TUBE (mm)	FOOT SPACING BETWEEN SUPPORTS	SPACING IN METERS (Approx.)
1/4" - 1/2"	6 - 13 mm	3 ft.	.9 m
3/8" - 7/8"	14 - 22 mm	4 ft.	1.2 m
1"	23 - 30 mm	5 ft.	1.5 m
1-1/4" & up	31 & up mm	7 ft.	2.1

Fig. 10.18- Guidelines for Tube Clamping Distances

36

36

❖ **Avoid Mechanical Stresses on Hard Transmission Lines:**

1. Avoid excessive stress on joints by selecting appropriate clamping spots.
2. Hard transmission lines shall not be used for equipment support.
3. Avoid poor tube bending.
4. Do NOT lay or stand on transmission lines during system servicing.

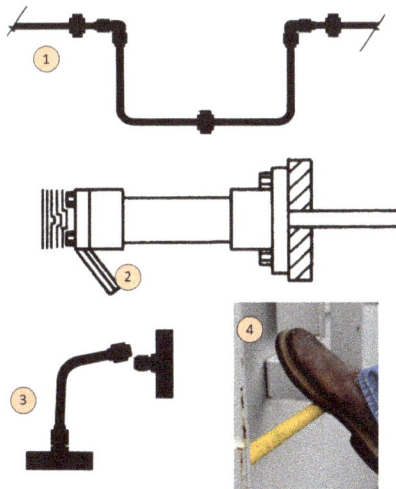

Fig. 10.19- Guidelines to Avoid Mechanical Stresses (Courtesy from Parker Hannifin)

37

37

10.4-BP-Transmission Lines-04-Standard Tests and Calibration

❖ **Standard:** ISO 6605 2002

- Hydraulic Hoses and Hose Assemblies-Test Procedures.

- This international standard specifies test methods for evaluating the performance of hoses and hose assemblies used in hydraulic fluid power systems. The following are common tests:

❖ **Pressure Proof Test (in accordance with ISO 1402):**

- Used to proof the normal specifications of a hose assembly.

- Is a nondestructive test.

- The hose assembly is hydrostatically tested with pressure.

- Period 30-60 seconds.

- No signs of leakage or failure → the assembly is considered "pass".

38

38

❖ **Burst Pressure Test (blog.parker.com):**
- Used to determines the actual burst strength of the assembly.
- This is a destructive test.
- The hose assembly is hydrostatically tested with pressure.
- No signs of leakage, bulging, coupling ejection or hose burst below the specified minimum rated burst pressure → the assembly "pass".
- Minimum burst values is based on :
  o Reasonable and safe maximum work pressure.
  o Safety factor (4-6).
  o Hose lifetime.

Video 275 (1 min)

Fig. 10.20- Hose Burst Test

39

39

❖ **Cold Flexibility Test (in accordance with ISO 10619):**

▪ Used to determine the ability of a hose to operate under cold weather.

▪ It is a destructive test.

▪ Condition hose assembly at T_{min} in a straight position for 24 hours.

▪ While still at the T_{min}, bend the sample on a mandrel over a time of 8-12 seconds as follows:

  ○ For hose (ID) sizes <= 22 mm bend them through 180^0 mandrel.

  ○ For hose (ID) sizes > 22 mm bend them through 90^0 mandrel.

▪ After bending, allow the sample to warm to room temperature.

▪ Visually inspect the sample for no hose or cover cracks.

▪ No leak when subjected to proof pressure → hose assembly "PASS".

40

40

❖ **Impulse Test:**

▪ Used to determine the ability of a 100Rxx hose to work under dynamic pressure.

▪ It is a nondestructive test.

▪ The hose is held in either a 90° or 180° configuration.

▪ The hose is exposed to cyclic pressure at 133% of working pressure.

▪ Frequency is 1 cycle/s.

▪ The hose meets (or exceed) the minimum number of impulse cycles → "PASS".

41

41

❖ **Salt Spray Test (ASTM B117):**

- It is also called Salt Fog Test.

- Used to check corrosion resistance of materials and surface coatings.

- It is a destructive test.

- Coated samples exposed to an accelerated corrosive attack.

- Visually evaluate and compare the suitability of the coating for use as a protective finish.

Fig. 10.21- Hose Burst Test (Courtesy from Gates)   42

42

Gravimetric Cleanliness Measurement (ISO 4405):

- Used to determine (for any hydraulic component):

  (total mass of contaminant inside a component/total inside surface area)

- It is a destructive test.

- A fluid is used to dislodge contamination in a hose assembly.

- The fluid is then poured through a membrane catch filter.

- The filtered particles are dried and weighed (in milligrams).

- Gravimetric cleanliness is calculated based on the total inside surface of the hose.

43

43

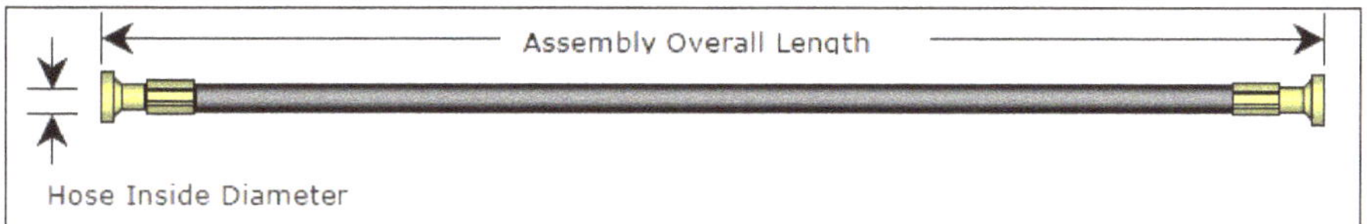

Hose Inside Diameter

Hose Diameter = -12 (3/4")
Assembly length = 246"
Surface Area (in) = Internal Circumference x Length
⇒ Internal Circumference = 3.1416 x Inside Diameter
3.1416 x 0.75"
2.36 in.

2.36 in. x 246 in.
580.6 in.^{2}

Surface Area (m) = 580.6 in.^{2} x (0.0006452)   *(0.0006452 converts in.^{2} to m^2)*
0.375 m^2

Contaminant Level = 52 mg / 0.375 m^2

138 milligram per square meter (mg/m^2)

Fig. 10.22- Hose Cleanliness Measurement (Courtesy from Gates)   44

44

❖ **Reliability Assessment Test (Fig. 10.23):**
▪ Majority of hoses from different manufacturers pass the regular impulse test → regular impulse test can't be used for comparative study.
▪ Comparative study requires a special test protocol.
○ Application specific parameters.
○ Test to failure or a specific number of cycles required by the application.
▪ The Hose Reliability Assessment is developed by (MSOE).

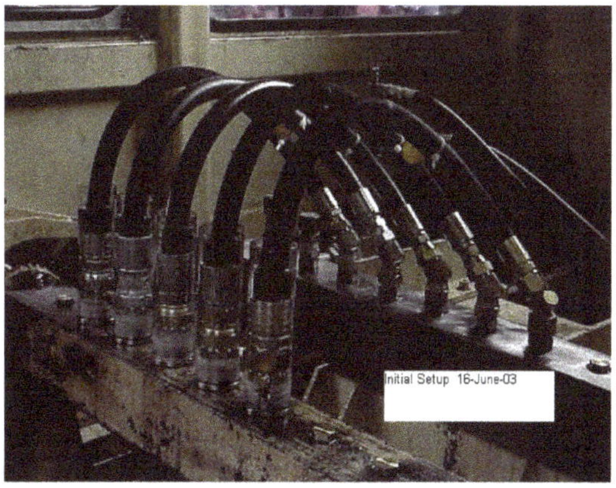

🎥 Video 137 (1 min)

Fig. 10.23- Hose Reliability Assessment Test

45

45

- **Test Specifications:**
  o Developed to assess the reliability of hoses used on refuse vehicles.
  o Manufacturers: 4 different hose manufacturers.
  o Hose Constructions: 2 different hose constructions (2W and 4W spiral).
  o Specimens: multiple specimens from each manufacturer and hose type.
  o Operating conditions: are based on vehicle measured data.
  o Operating conditions: (pressure, temperature, and pulsation frequency).
  o Bend Radius: 1/2 of the minimum specified by hose construction.
  o Twisting Angle: 45 Degrees between hose couplings.
  o Minimum Test Cycles: 2,000,000 cycles.

- **Test Results:**
  o 4W spiral hose lasts up to 5 times longer than the 2W hose.
  o Increased vehicle productivity.
  o Reduced environmental costs of cleanups.
  o Estimated company annual savings was $10M.

46

46

10.5-BP-Transmission Lines-05-Transportation and Storage

- **Hose Storage (1):** coiled on reels that rotate manually or using a motor.
- **Tube/Pipe Storage (2):**
  o They should be placed on supports.
  o Distances are based on the size and length to avoid bending.
  o Pipe manufacturers must be reviewed for such information.
- **Dust Caps (3):** All lines ends should be closed by proper dust caps.

Fig. 10.24- Best Practices for Storage of Transmission Lines

47

47

Video 151 (2 min)

- **Storage Time:**
  - o Maintain a system of age control (first-in first-out).
  - o A hose must be used before the shelf life is expired.
  - o Follow the manufacturer's shelf life.
  - o However, SAE J517 specifies the shelf life of hydraulic hoses.

- **Transportation:**
  - o Transmission lines of all types must be sealed with protective caps.

48

48

Chapter 10 Reviews

1. When replacing a transmission line by other one of different size, if the flow rate through the line is kept the same, which of the following operation conditions is affected?
   A. Flow pattern through the valve (laminar/turbulent).
   B. Fluid speed inside the line.
   C. Pressure drop across the line.
   D. All of the above.

2. Which of the following cleaning process is considered last cleaning before use?
   A. Pickling.
   B. Flushing.
   C. Projectile cleaning.
   D. Cleaning by water.

3. When a hydraulic line is leaking, which of the following consequences is most critical and require immediate response
   A. Cost of the fluid.
   B. Possible oil injection through an operator body.
   C. Environmental pollution.
   D. Loss of pressure.

4. Using clamps to hold transmission lines firmly in a machine ir required to?
   A. Reduced noise and vibration of the lines.
   B. Increase allowable working pressure inside line.
   C. Reduce pressure drop across the line.
   D. All of the above.

5. In a hydraulic line, 50 mg of contaminates were found. If the line inside surface is 0.2 square meter, the gravimetric cleanliness level is?
   A. 50 mg
   B. 500 mg/m^2
   C. 250 mg
   D. 250 mg/m^2

Chapter 10 Assignment

Student Name: --- Student ID: ------------------

Date: -- Score: -----------------------

Question: List the cases when a machine shut down and line replacement are required, no matter the line life is

**Chapter 11
Maintenance of Heat Exchangers**

Objectives:

This chapter provides guidelines for **heat exchangers** selection, replacement, maintenance scheduling, installation, testing, storage and transportation. This chapter is supported by examples and figures granted by leading fluid power manufacturers.

Brief Contents:

11.1-BP-Heat Exchangers-01-Selection and Replacement

11.2-BP-Heat Exchangers-02-Maintenance Scheduling

11.3-BP-Heat Exchangers-03-Installation and Maintenance

11.4-BP-Heat Exchangers-04-Standard Tests and Calibration

0

0

The following topics are discussed in Chapter 5 in Volume 4 "Hydraulic Fluids Conditioning" of this series of textbooks:
- Contribution of Heat Exchangers:
- Air-Type versus Water-Type Oil Coolers
- Determination of Cooling Capacity for an Oil Cooler
- Air-Type Oil Coolers
- Shell-and-Tube Water-Type Oil Coolers
- Plat-Type Oil Coolers
- Cooling-Filtration Units
- Oil Cooling Circuit Diagram
- Oil Temperature Automatic Control Solutions
- Electrical Oil Heaters

The following topics are discussed in Chapter 11 in Volume 6 "Troubleshooting and Failure Analysis" of this series of textbooks:
- Heat Exchangers Inspection
- Heat Exchangers Troubleshooting
- Heat Exchangers Failure Analysis

1

1

11.1-BP-Heat Exchangers-01-Selection and Replacement

- Heat Exchangers are specified based on cooling or heating capacity.

- When replacing an existed heat exchanger, size is the main design parameter.

- Improperly sized heat exchanger →:
  - Hydraulic system temperature instability.
  - All consequences of hydraulic fluid viscosity increasing/decreasing.
  - Possible pump cavitation.
  - Reduced system reliability.

2

2

11.2-BP-Heat Exchangers -02-Maintenance Scheduling

Unless otherwise stated by components and systems manufacturer:

#	Preventive Maintenance Actions	Daily	Weekly	Monthly	Biannually	Annually
1	Check for external leaks	✔	✔	✔	✔	✔
2	Clean the dust on outer surfaces		✔	✔	✔	✔
3	Check hydraulic connections			✔	✔	✔
4	Check coolant connection			✔	✔	✔
5	Check proper connection with electrical instrumentation			✔	✔	✔
6	Check Zinc Anodes for corrosion **(Note 1)**			✔	✔	✔
7	Inside deep cleaning **(Note 2)**				✔	✔
8	Check proper setting of temp. control system **(Note 3)**				✔	✔
9	Check for internal leaks					✔

Table 11.1- BP-Heat Exchangers-02-Maintenance Scheduling

3

3

❖ **Note 1 (Zinc Anodes):**
- Cooling water may contain mineral salts →
- Corrosion of connections between the cooler and water supply lines.
- → Zinc Anodes protect the heat exchanger material.
- → Frequent inspection is required.

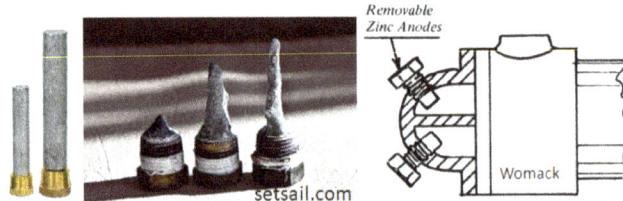

Fig. 11.1- New and Used Zinc Anodes

❖ **Note 2 (Deep Cleaning):**
- Cleaning tube and shell of a water-cooled heat exchanger.
- Cleaning fins of an air-cooled heat exchanger more frequently.

❖ **Note 3 (Temperature Setting):**
- Proper setting of thermostat for either On/Off or analog temperature control system.

4

4

11.3-BP-Heat Exchangers-03-Installation and Maintenance

Heat exchanger failure → complete shutdown of operations.
Routine maintenance → avoid unplanned shutdowns.

❖ **Proper Placement of a Water-Type Heat Exchangers:**
- Should be placed in a well-ventilated area.
- Should be placed apart or shielded from external heat sources.
- Should be shielded from harsh weather (sunlight or freezing temperature).
- Should be protected from moving machinery (swinging booms, lift trucks).
- Install heat exchangers horizontally → gravity of oil and the coolant won't affect the speed of circulation. However, requirements for draining the circuits should be considered.

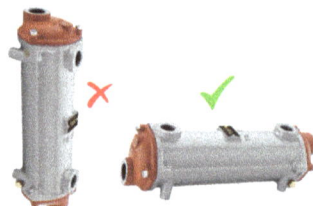

Fig. 11.2- Correct Orientation when Installing Water-Cooled Heat Exchangers

5

5

❖ **Proper Filter-Cooler Assembly:**
▪ Capturing oil contaminants is easier when the oil is hot.
▪ Pressure drop across the filter is less when the oil is hot.
▪ If cold oil is forced into a filter, back pressure may damage the tubes in the cooler.
▪ Forcing oil into a cooler before filtering it reduces the cooler efficiency as dirt can accumulate inside the heat exchanger.

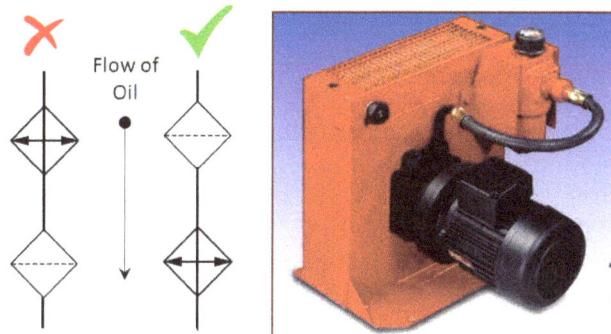

Fig. 11.3- Proper Filter-Cooler Assembly

6

6

❖ **Fouling of Water-Type Oil Cooler:**
▪ Fouling is deposition of any undesired material on a cooler surfaces.
▪ Fouling → significantly impact the thermal performance of coolers.
▪ Fouling → obstruct fluid flow, corrosion ↑ pressure drop ↑.

▪ **Corrosion Fouling (1):**
  o Chemical reaction (rust) on heat exchanger surface material:
  o Even thin coatings of oxides → significantly affect thermal performance.

▪ **Particulate/Sedimentation Fouling (2):**
  o Particles (dirt, sand etc.) deposited on the heat transfer surface.
  o Some of these deposits are difficult to remove mechanically.

Fig. 11.4- Fouling of Heat Exchangers (hcheattransfer.com/fouling1. html)

7

7

- **Scaling/Crystallization Fouling (3):**
  - Due to salts such as calcium carbonate ($CaCO_3$) found in water.
  - Scale is difficult to remove mechanically.
  - Chemical cleaning may be required.

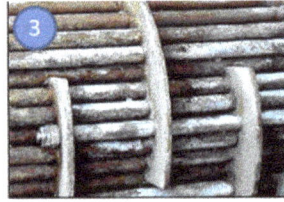

- **Biological Fouling (4):**
  - Untreated water → Biological microbes growth on heat transfer surfaces.
  - → affect thermal performance.

- **Varnish Fouling (5):**
  - Increased working temperature →
  - oil degradation → varnish formation

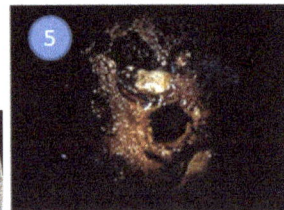

8

8

- **Salt Fouling (6):** Occurs as a result of salt depositing if sea water is directly used without desalination.

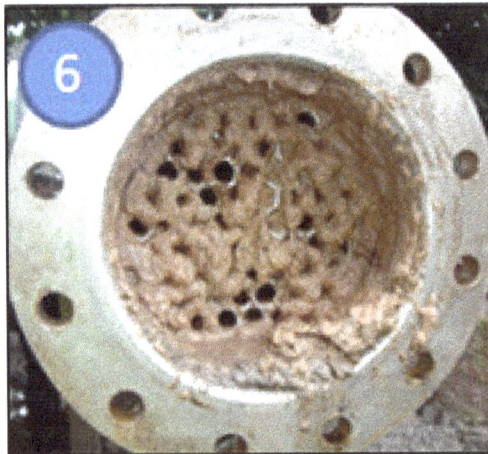

Video 438 (1 min)

9

9

❖ **Cleaning of Water-Cooled Heat Exchangers:**
- Routine cleaning of tube and shell →
  - Reduces water usage.
  - Reduce unexpected shutdowns.
  - Improves heat exchanger efficiency.
  - Improve system reliability and operation cost effectiveness.

- Clean oil is flowing through the shell → Clean only inside the water tubes.
- Varnish has built up in oil →
  - Clean outside and inside water tubes.
  - Clean inside surfaces of the shell.

10

10

Small Heat Exchangers:
- For small or lightly contaminated heat exchangers → **Traditional Flushing** → circulating hot wash oil through the tube and shell → remove sludge or similar soft deposits.

- Soft salt deposits → wash out by circulating fresh hot water or a mild alkaline solution (1.5% solution of sodium hydroxide or nitric acid)

- **Note:** Tubes should be
- Flushed in the opposite direction of the normal oil flow.
- Dried after flushing by hot and dry air to remove flushing liquid residuals.

11

11

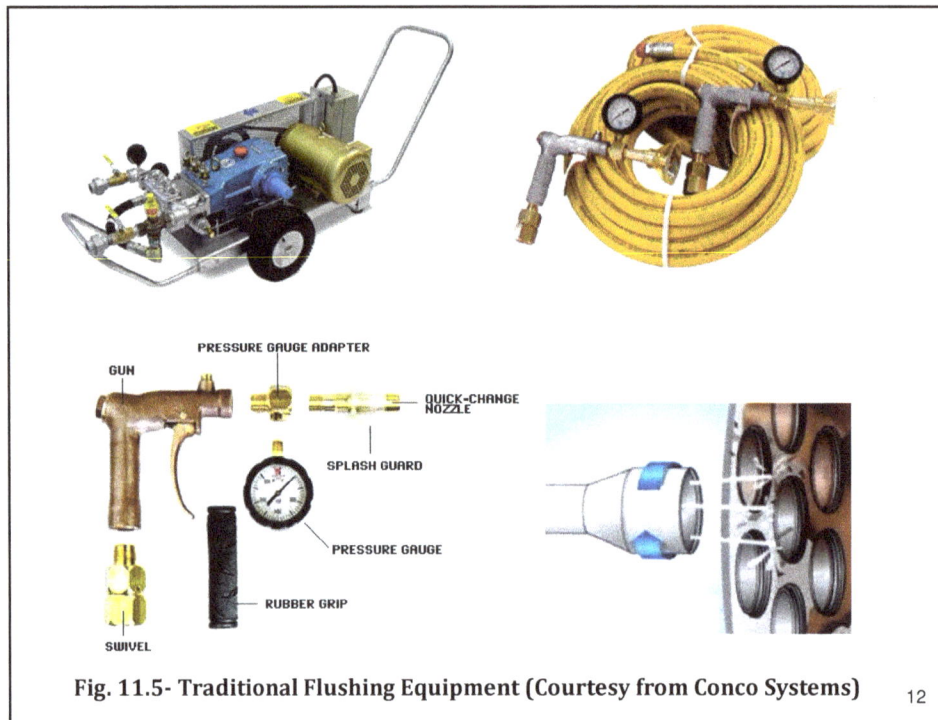

Fig. 11.5- Traditional Flushing Equipment (Courtesy from Conco Systems) 12

12

Large Heat Exchangers:

- Use **Power Flushing** systems → push cleaning plugs through the tube → removing all types of fouling.
- Plug speed (10 – 20) ft/s at pressure (200-300) PSI.
- DO NOT use AIR to push the cleaning plugs
- Use water to provides lubrication.
- Depending on the type of fouling → one or combination of the following techniques shall be followed.

13

13

Cleaning Straight Tubes (1) and U-Tubes (2) by Plastic Tube Cleaner (1):
- Remove only soft deposits (mud, silt and microbes).
- Available in sizes 5/8" to 1-1/4" I.D.

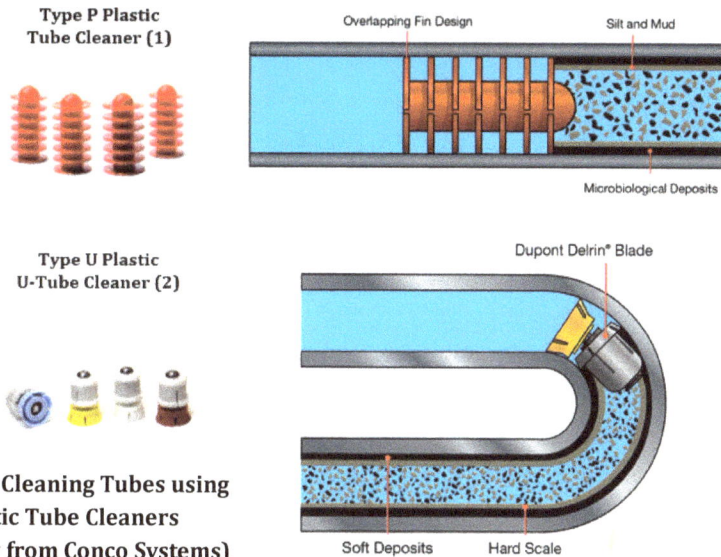

Type P Plastic Tube Cleaner (1)

Overlapping Fin Design   Silt and Mud

Microbiological Deposits

Type U Plastic U-Tube Cleaner (2)

Dupont Delrin® Blade

Fig. 11.6- Cleaning Tubes using Plastic Tube Cleaners (Courtesy from Conco Systems)

Soft Deposits   Hard Scale

14

14

Cleaning Tubes by Plastic H-Brush Tube Cleaner (1):
- Remove light deposits, micro/macro fouling, soft organic deposits, some corrosion by-products, mud and silt, and most types of obstructions.
- Available in sizes 5/8" to 1-1/2" I.D.

Cleaning Tubes by Plastic XL-Brush Tube Cleaner (2):
- Similar characteristics to the H-Brush but longer

Cleaning Tubes by Stainless Steel Tube Cleaning Brush (3):
- Remove iron, silica, obstructions, macrofouling and debris.
- Used for all types of tubs including inserts and epoxy coatings.
- Available in sizes 5/8" to 1-1/4" I.D. ,
- Restore the tube surface to its original heat transfer characteristics.

15

15

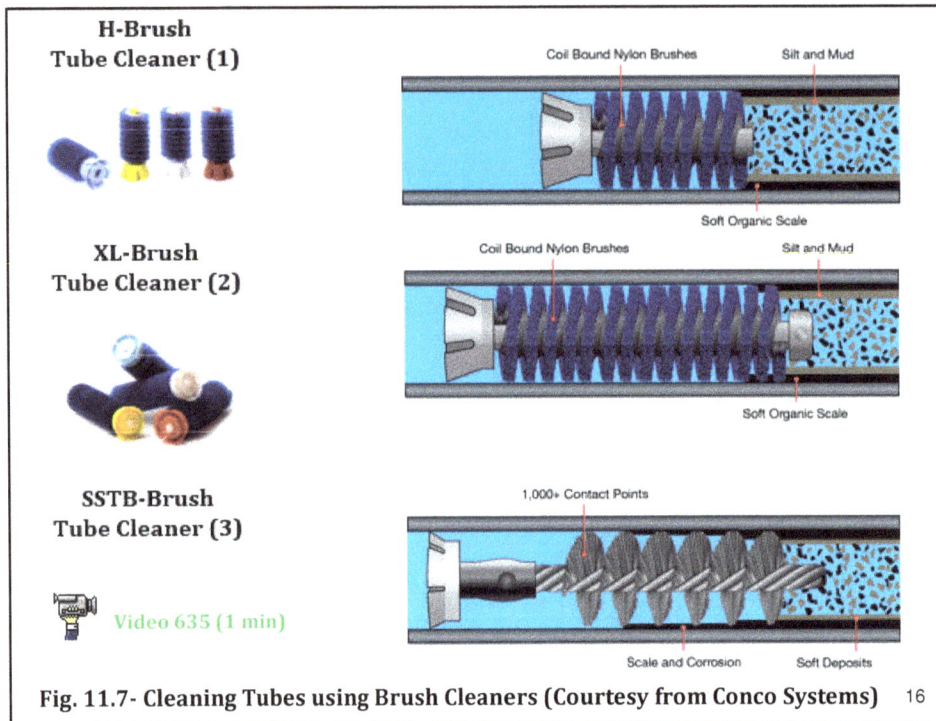

H-Brush Tube Cleaner (1)

XL-Brush Tube Cleaner (2)

SSTB-Brush Tube Cleaner (3)

Video 635 (1 min)

Fig. 11.7- Cleaning Tubes using Brush Cleaners (Courtesy from Conco Systems)   16

16

Cleaning Straight Tubes by **C2x Metallic Tube Cleaner (1):**

- **Two-stage** design with **six points of cleaning contact** per blade.
- Available in **sizes** ¾", 7/8", 1" and 1-1/8" I.D. .
- **Remove** thin layers of silica, iron and calcium, corrosion, debris and obstructions.
- For tubes that are **extremely fouled,** consider the C3X tube cleaner featuring a **three-stage design.**

Cleaning **U-Tubes by Metallic Tube Cleaner (2):**

- **Navigate** in U-tubes.
- Remove deposits that softer cleaners can't touch.
- Available in **sizes** 5/8" to 1-1/4" I.D.

17

17

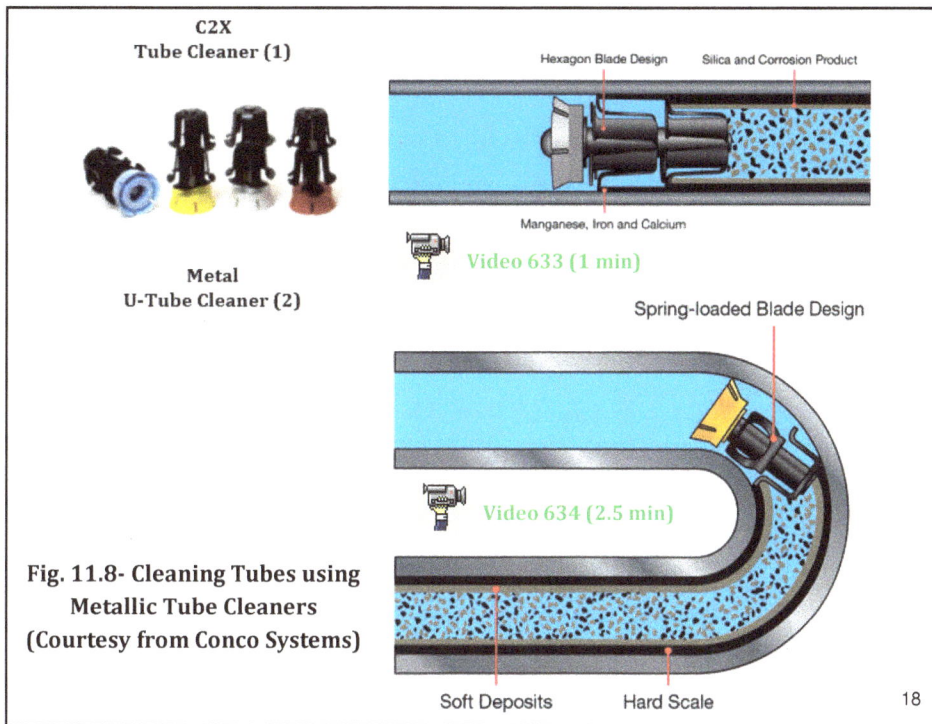

Fig. 11.8- Cleaning Tubes using Metallic Tube Cleaners (Courtesy from Conco Systems)

18

❖ **Cleaning of Air-Cooled Heat Exchangers:**

▪ Use brush, compressed air, or by steam.

▪ Remove dust, dirt, debris, pollen, leaves and other deposits.

▪ Remove dirt which may stick between the fins/tubes.

Video 636 (1 min)   Video 167 (1 min)

Fig. 11.9- Cleaning Radiators of Air-Cooled Heat Exchangers

19

11.4-BP-Heat Exchangers-04-Standard Tests and Calibration

Test for Internal Leaks:

- Corrosion or pressure exceeds specification →
- Internal leaks can occur between the oil and water chambers.
- Routinely inspect sample of the oil and water.
- Cooling tower → withdraw the sample from the top of the tower.

Video 675 (1 min)

Fig. 11.10- Pressure Test of Heat Exchangers

20

20

Chapter 11 Reviews

1. Among the maintenance actions of the heat exchanger, what is the most frequent action?
    A. Check for external leakage.
    B. Check hydraulic connections.
    C. Inside deep cleaning.
    D. Check for internal leakage.

2. Zink Anodes are used for?
    A. Improve thermal properties of air-cooled heat exchangers.
    B. Improve thermal properties of water-cooled heat exchangers.
    C. Protect material of air-cooled heat exchangers.
    D. Protect material of water-cooled heat exchangers

3. A filter must be placed before a cooler because of:
    A. Capturing oil contaminants is easier when the oil is hot.
    B. Pressure drop across the filter is less when the oil is hot.
    C. If cold oil is forced into a filter, back pressure may damage the tubes in the cooler.
    D. All of the above.

4. Varnish in tubes means that?
    A. Biological microbes growing in the heat exchangers.
    B. Oxidation and rusting due to water content.
    C. Oil degradation due to overheating.
    D. Presence of salt in water.

5. A water-type heat exchanger should be placed horizontally because, in that position:
    A. The heat exchange will be well ventilated.
    B. The heat exchanger is protected from external heat.
    C. Gravity of oil and coolant won't affect the speed of circulation.
    D. The heat exchanger is protected from moving surfaces.

Chapter 11 Assignment

Student Name: -- Student ID: -------------------

Date: --- Score: ------------------------

Question: Discuss best practices for water-type heat exchangers.

**Chapter 12
Maintenance of Filters**

Objectives:

This chapter provides guidelines for **Filters** selection, replacement, maintenance scheduling, installation, testing, storage and transportation. This chapter is supported by examples and figures granted by leading fluid power manufacturers.

Brief Contents:

12.1-BP-Filters-01-Selection and Replacement
12.2-BP-Filters-02-Maintenance Scheduling
12.3-BP-Filters-03-Installation and Maintenance
12.4-BP-Filters-04-Standard Tests and Calibration
12.5-BP-Filters-05-Transportation and Storage

0

0

The following topics are discussed in Volume 3
"Hydraulic Fluids and Contamination Control"
- Chapter 2: Hydraulic Fluids
- Chapter 3: Energetic Contamination
- Chapter 4: Gaseous Contamination
- Chapter 5: Fluidic Contamination
- Chapter 6: Chemical Contamination
- Chapter 7: Particulate Contamination
- Chapter 8: Hydraulic Fluid Analysis
- Chapter 9: Hydraulic Filters Performance Ratings
- Chapter 10: Contamination Control in Hydraulic Transmission Lines

The following topics are discussed in Volume 4
"Hydraulic Fluids Conditioning":
- Chapter 01: Introduction to Hydraulic Filters
- Chapter 02: Filter Media and Filtration Mechanisms
- Chapter 08: Filter Selection Criteria

1

1

The following topics are discussed in Chapter 12 in Volume 6
"Troubleshooting and Failure Analysis" of this series of textbooks:

- Filters Inspection

- Filters Troubleshooting

- Filters Failure Analysis

2

2

12.1-BP-Filters-01-Selection and Replacement

- **Filters are originally specified based on:**
  - Cleanliness level prescribed by the system designer.
  - Placement in the circuit.
  - Size and dirt holding capacity.
  - Static and dynamic working conditions (P, T, Q)
  - Mechanical mounting method.

- **When replacing an existed filter:**
  - None of the originally specified design and operating specifications shall be changed.

3

3

Importance of maintaining the specification of a filter.

- Example:
- Amount of dirt (passes through a pump) vs. oil cleanliness level.

- Test Conditions:
- 200 lit/min flow, 18 hours a day, and 340 working days per year).
- Oil is typically contaminated with particles to ISO 19/17/14.

ISO Code	NAS 1638	Description	Suitable for	Dirt/year
ISO 14/12/10	NAS 3	Very clean oil	All oil systems	7.5 kg *
ISO 16/14/11	NAS 5	Clean oil	Servo & high pressure hydraulics	17 kg *
ISO 17/15/12	NAS 6	Light contaminated oil	Standard hydraulic & lube oil systems	36 kg *
ISO 19/17/14	NAS 8	New oil	Medium to low pressure systems	144 kg *
ISO 22/20/17	NAS 11	Very contaminated oil	Not suitable for oil systems	> 589 kg *

Table 12.1- Amount of Dirt Pass through a Filter based on Oil Cleanliness Level (Courtesy of C.C. Jensen Inc.)

4

4

12.2-BP-Filters-02-Maintenance Scheduling

Unless otherwise is the stated by components and systems manufacturer:

#	Preventive Maintenance Actions	Daily	Weekly	Monthly	Biannually	Annually
1	Clean the dust on outer surface		✔	✔	✔	✔
2	Check hydraulic connections			✔	✔	✔
3	Check status of the filter through clogging indicator **(Note 1)**		✔	✔	✔	✔
4	Check electrical connections (if found)			✔	✔	✔
5	Disassemble and inspect/clean/wash/replace filter element and filter housing **(Note 2)**			✔	✔	✔
6	Check valve performance through standard tests				✔	✔

Table 12.2- BP-Filters-02-Maintenance Scheduling

5

5

Note (1) – Check Filter Status:

- Fitter status → health of the overall system.

- Clogging indicators should be readily visible to the operator.

- No clogging indicators → scheduled routine filter element replacement.

Note (2): - Disassembling and Replacement:

Take a good look and check if the filter has any of the following signs:

- Damage to the filter element pleats (cuts or bunched).

- Filter elements sealing on both end caps.

- Center tube is collapsed or buckled.

- Oil degradation products accumulated.

- High concentration of debris as a sign of metal wear.

- Nonmetallic particles such as paint chips, fibers, seal wear products, etc.

6

6

12.3-BP-Filters-03-Installation and Maintenance

- Good maintenance practices → consistent filtration performance.
- Contamination induced by a filter change → goes directly to the system.
- Example of best practices:

Review available instruction (1):

- Best source of instructions → system designer and filter manufacturer.
- Review filter Pictogram.

Check the service indicator (2):

- Verify that service indicators show service interval has been reached.

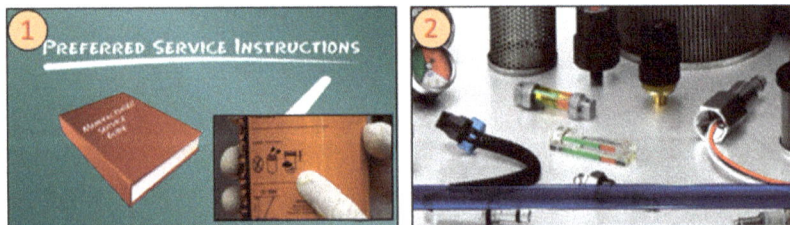

Fig. 12.1- Hydraulic Spin-On Filter Replacement Steps (Courtesy of Donaldson)   7

7

Turn off system pressure (3):

- Be sure the system is turned off and release any residual pressure.
- Isolate the filter under service (if isolation setup is found).

Remove the used filter and gasket (4):

- Remove the filter,
- Dispose of the filter in accordance with local regulations.

8

8

Clean the filter mounting head and bowel (5):

- Use only lint free wipes or filtered air to dry the bowl.
- Clean the surfaces of the filter head or cover.
- Flush sediments from filter bowls (pre-filtered solvent).

Lubricate the filter gasket (6):

- Lubricate threads and spin on seal with clean system oil.

9

9

Inspect new filter (7):
- DO NOT install any filter or filter element that shows any signs of damage.
- Clean exterior of the filter housing.
- Remove the plastic bag only in place.
- Avoid touching a new element if possible.
- After replacing the new filter element, secure the bowl immediately.

Install new filter for instructions (8):
- Install the filter until the gasket first contacts the sealing surface.
- Final tightening in accordance with the given instructions.

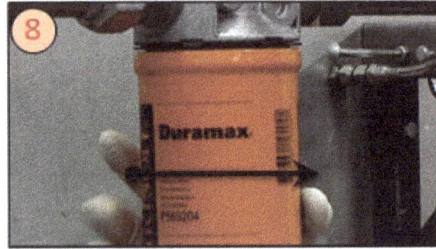

10

10

Video 400 (7 min)

Fig. 12.2- Spin-On
Hydraulic Filters
Service Pictograms
(Courtesy from
Donaldson)

11

11

12.4-BP-Filters-04-Standard Tests and Calibration

❖ **Standard Test Methods → Performance Ratings of Filter:**

- **ISO 2942:** Filter Element Structural Integrity (Bubble Point) Test.

- **ISO 2943:** Hydraulic Fluid Compatibility Test

- **ISO 16889:** Efficiency and Capacity (Multipaas) Test

- **ISO 3723-2015:** End Load Test

- **ISO 3968:** Differential Pressure Test

- **NFPA (T-2.6.1):** Rated Burst Pressure (RBP) of a Filter Housing

- **ISO 1077-1:** Rated Fatigue Pressure (RFP) of a Filter Housing

- **Cyclic Test Pressure** (CTP) of a Filter Housing

- **ISO 2941:** Collapse Pressure of a Filter Element

- **ISO 3724 OR ISO 23181:** Flow Fatigue Test for Filter Element

12

12

❖ **Proposed Logic Sequence of Tests:**

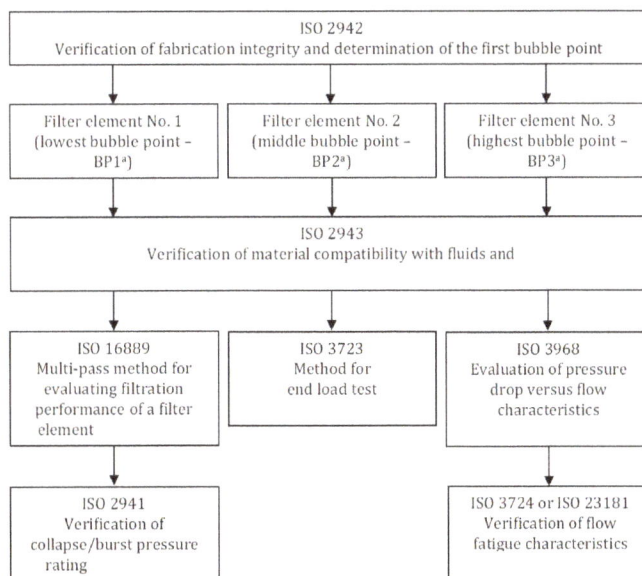

Fig. 12.3- Sequence of Conducting Standard Tests for Hydraulic Filters (Courtesy from Donaldson)

13

13

12.4.1- ISO 2942: Filter Element Structural Integrity (Bubble Point) Test.

❖ **Test Purpose is to determine:**

- Filter element structure acceptability for use or testing.
- Filter element meets prescribed maximum allowable pore size.
- No damage during shipping or manufacturing.

❖ **Test Method:**

- Bubble Point Resistance by checking the absence of bubbles.
- Submerging the test element in isopropyl alcohol, or other suitable fluid.
- Apply air pressure to the inside of the element.
- No evidence of a steady stream of bubbles is detected →
- The element is considered PASS the test.

14

14

12.4.2- ISO 2943: Hydraulic Fluid Compatibility Test

❖ **Test Purpose is to determine:**
- Compatibility with the hydraulic fluids at maximum temperature.

❖ **Test Method:**

- Submerge filter elements in the specified fluid.
- Test temperature 15° C (59° F) > the maximum operating temperature.
- Test Period: 72-hour period.
- No visual evidence of structural failure or material degradation →
- The element is considered PASS the test.

15

15

12.4.3- ISO 16889: Efficiency and Capacity (Multipaas) Test

12.4.3.1- Multipass Test Purpose and Procedure

❖ **Test Purpose is to determine:**

▪ The Multipass Test is the worldwide recognized method of characterizing hydraulic filter element filtration performance including efficiencies and dirt capacity.

❖ **Test Method:**

▪ Test fluid (Mil-H-5606) is injected by uniform contaminants.

▪ Test contaminants (ISO 12103-A3 OR MTD, ISO Medium Test Dust).

▪ Pumped the contaminated fluid through test filter.

▪ Use automatic particle counter.

▪ Count particulate contaminants of specific sizes at upstream side (Nu).

▪ Count particulate contaminants of specific sizes at downstream side (Nd).

▪ Obtain the beta ratio.

16

16

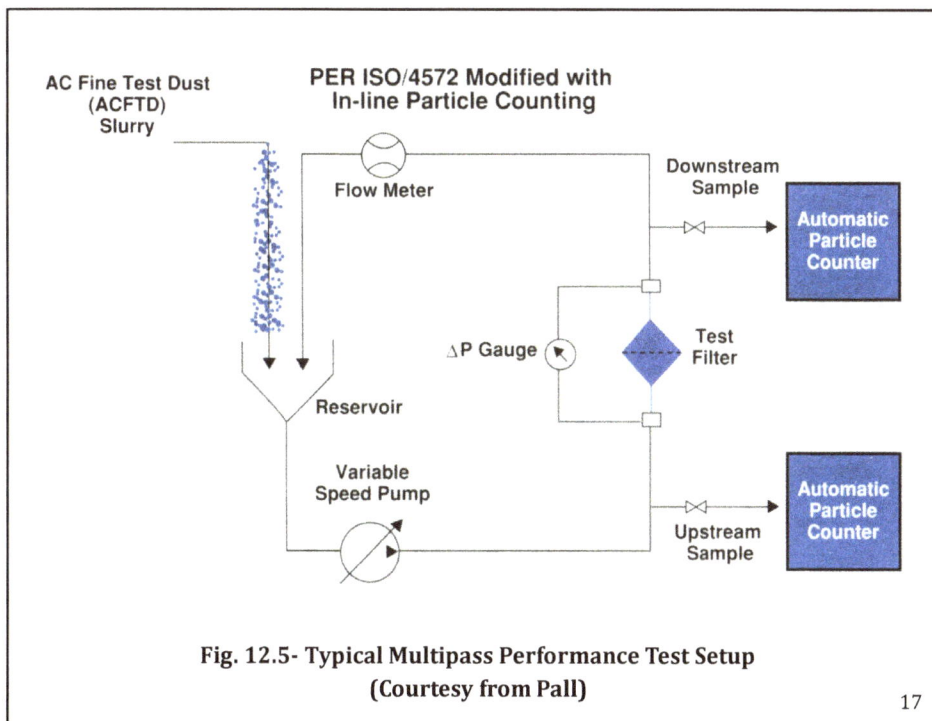

**Fig. 12.5- Typical Multipass Performance Test Setup
(Courtesy from Pall)**

17

17

12.4.3.2- Calculation of Beta Ratio

$$\beta_x = \frac{N_U}{N_D} \qquad\qquad 12.1$$

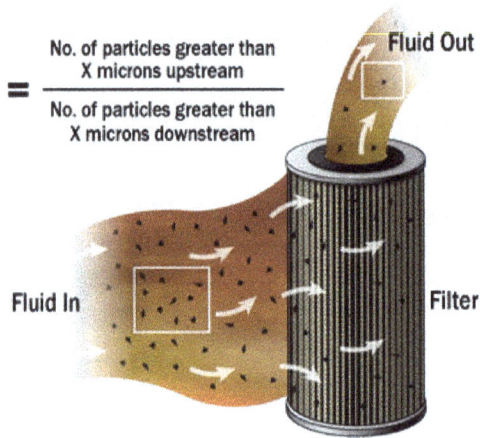

$$\beta x = \frac{\text{No. of particles greater than}}{\text{No. of particles greater than}}$$
X microns upstream / X microns downstream

Fluid Out

Fluid In

Filter

Fig. 12.6- Calculation of Beta Ratio (www.magneticfiltration.com) [18]

18

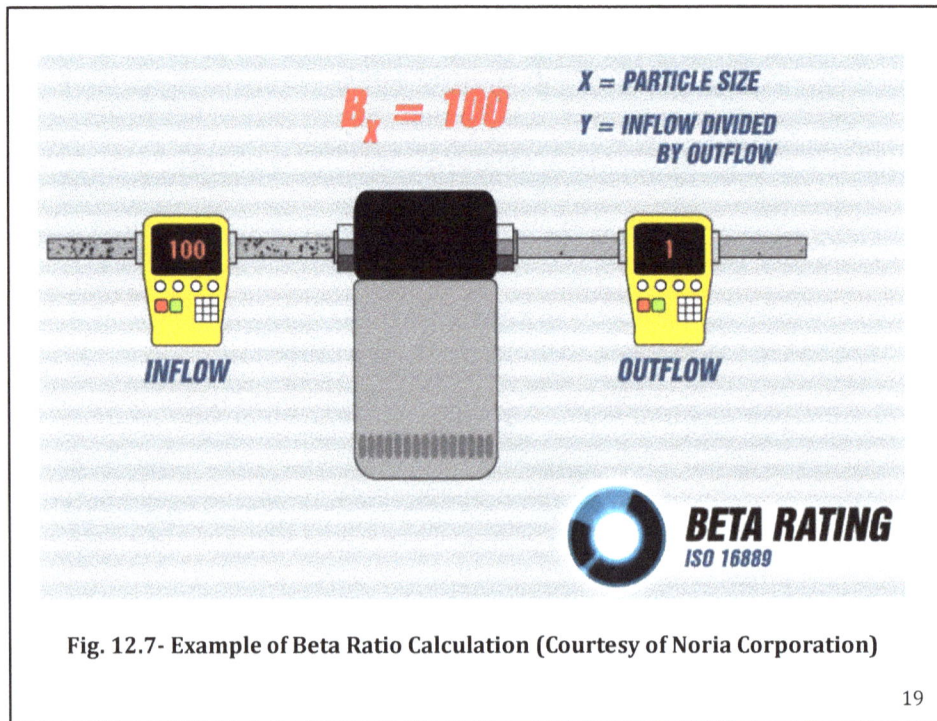

$B_x = 100$

X = PARTICLE SIZE
Y = INFLOW DIVIDED BY OUTFLOW

100

1

INFLOW

OUTFLOW

BETA RATING
ISO 16889

Fig. 12.7- Example of Beta Ratio Calculation (Courtesy of Noria Corporation)

19

Interpretation of Typical data from a Multipass Test:
- Beta ratio =12 for particle size > 2 μm,
- Beta ratio = 100 for particle size > 5 μm,
- Beta ratio = 3000 for particle size > 10 μm.

Particle Size (μm)	Particle Counts (#/ml)		Beta Ratio
2	upstream downstream	15,200 1,267	$\beta_2 = 12$
5	upstream downstream	8,000 80	$\beta_5 = 100$
10	upstream downstream	3,000 1	$\beta_{10} = 3000$

Table 12.3- Typical Multipass Test Data (Courtesy of Pall)

20

20

Effect of Particle Size Service Life:
- Better filter rating → Longer bearing service life.

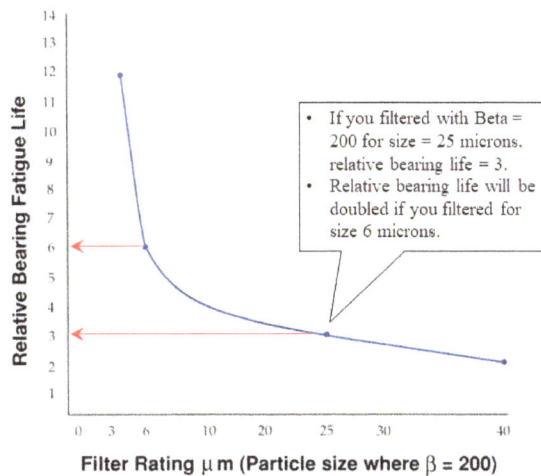

- If you filtered with Beta = 200 for size = 25 microns, relative bearing life = 3.
- Relative bearing life will be doubled if you filtered for size 6 microns.

Filter Rating μ m (Particle size where β = 200)

Ref: Macpherson, P.B., Bhachu, R., Sayles, R., "The Influence of Filtration on Rolling Element Bearing Life"

Fig. 12.8- Effect of Beta Ratio on Bearing Life

21

12.4.3.3- Beta Ratio Stability

- The Multipass test is performed under controlled laboratory conditions.
- Varying operating conditions (pressure, flow, temperature, vibration)
- → unstable beta ratio →
- Beta ratio of a filter should be defined within range of worki
- Stable Beta Ratio → stable filter perform over time.

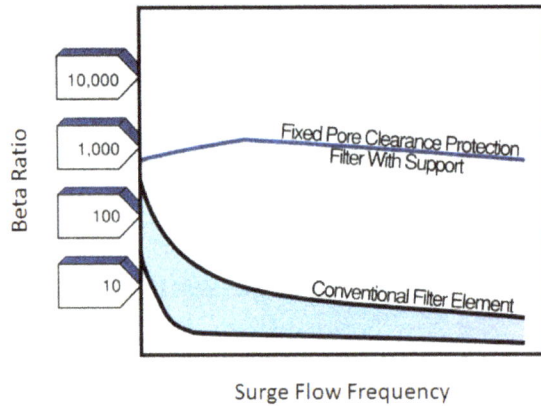

Fig. 12.9- Effect of Surge Flow on Beta Ratio (Courtesy of Pall)

22

12.4.3.4- Filter Efficiency

$$E_x = \left[1 - \frac{N_D}{N_U}\right] \times 100 = \left[1 - \frac{1}{\beta_x}\right] \times 100 = \left[\frac{\beta_x - 1}{\beta_x}\right] \times 100 \qquad 12.2$$

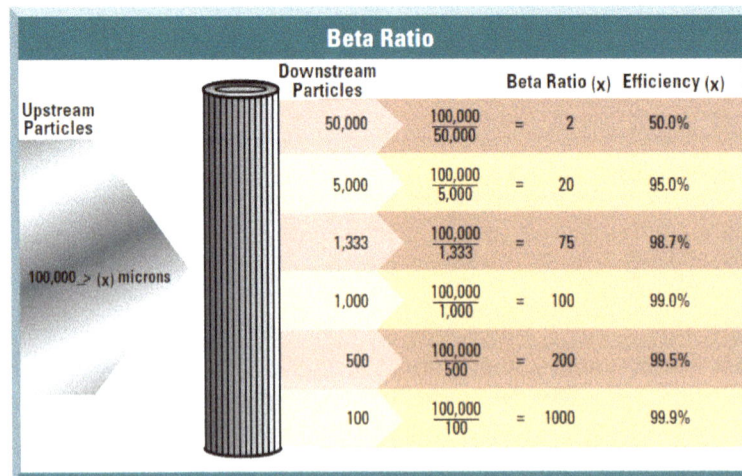

Beta Ratio				
	Downstream Particles		Beta Ratio (x)	Efficiency (x)
Upstream Particles	50,000	$\frac{100,000}{50,000}$ =	2	50.0%
	5,000	$\frac{100,000}{5,000}$ =	20	95.0%
	1,333	$\frac{100,000}{1,333}$ =	75	98.7%
100,000 > (x) microns	1,000	$\frac{100,000}{1,000}$ =	100	99.0%
	500	$\frac{100,000}{500}$ =	200	99.5%
	100	$\frac{100,000}{100}$ =	1000	99.9%

Fig. 12.10- Filter Efficiency vs. Beta Ratio (Courtesy of Parker)

23

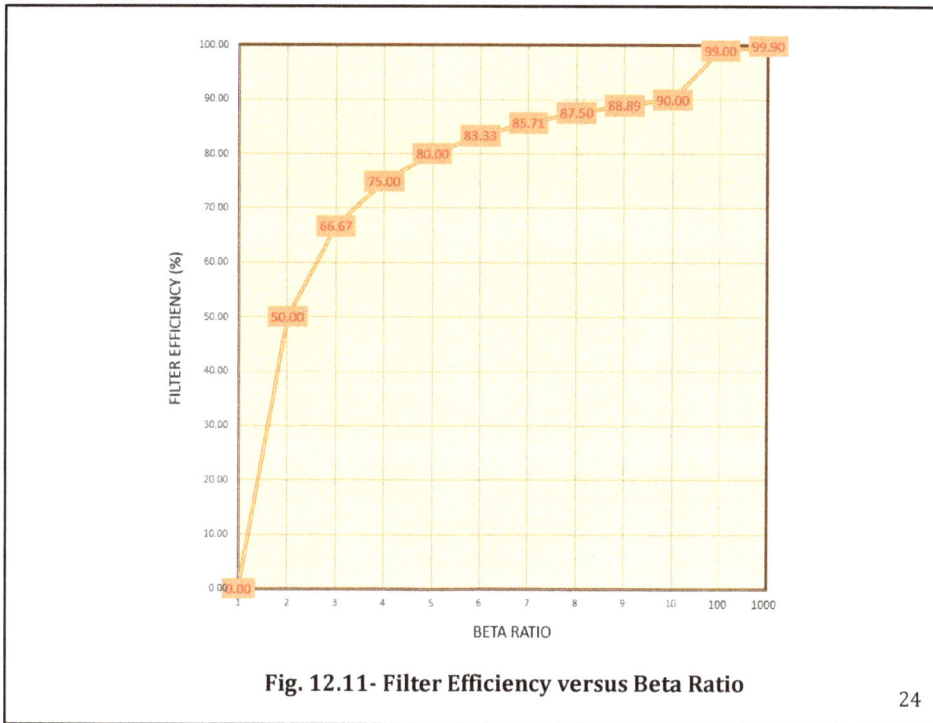

Fig. 12.11- Filter Efficiency versus Beta Ratio

24

24

12.4.3.5- Nominal and Absolute Ratings

- **Nominal Rating:** $\beta_X = 2$ ($E_X = 50$ %).
- **Absolute Rating:** $\beta_X = 75$ ($E_X = 98.7$ %).
- Filters can be rated for various particle sizes as B3/6/15 = 2, 10, 75.
- Filter is nominal at 3 microns.
- Filter is 90% efficient at 6 microns.
- Filter is absolute for 15 microns.

Filtration Ratio (at a given particle size)	Capture Efficiency (at the same particle size)
2	Nominal → 50 %
5	80%
10	90%
20	95%
75	Absolute → 98.7 %
100	99%
200	99.5%
1000	99.9%

Table 12.4- Nominal and Absolute Ratings

25

25

12.4.3.6- Filter Dirt Holding Capacity

Video 191 (0.5 min)

- Dirt Holding Capacity (DHC).
- It is the weight of dirt a filter can hold before ΔP = saturation.
- ISO 16889 → ISO MTD Test Dust is added → at specified ΔP →
- Measure DHC (total grams of dirt that a filter held).

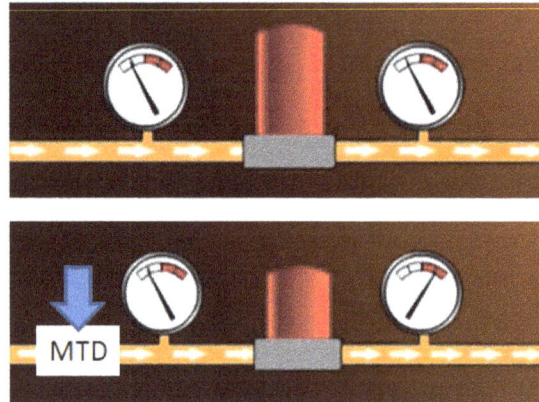

Fig. 12.12- Dirt Holding Capacity Test (Courtesy of Parker)

26

26

- Impact of DHC on Operation Cost:
- Filter DHC ↑ → Filter element replacement ↓ → cost of operation ↓

Cost of Removing 1 kg or lb of Dirt

$$= \frac{\textbf{Cost of Filter Element (Installation \& Disposal)}}{\textbf{Dirt Holding Capacity in kg or lb}} \qquad \textbf{12.3}$$

	Example 1	Example 2
Filter type	Glass fiber based pressure filter insert	Cellulose based offline filter insert
Cost of element/insert	€ 35 / $ 50	€ 200 / $ 300
Dirt holding capacity	0.085 kg / 0.18 lbs	4 kg / 8 lbs
Cost per kg/lb removed dirt	€ 412 / $ 278	€ 50 / $ 40

Table 12.5- Cost of Removing Dirt (Courtesy of C.C. Jensen Inc.)

27

27

- Example of stacked disc filter element:
  - 3 μm nominal → 50% of all particles larger than 3 μm are retained.
  - 8 μm absolute → 98.7% of all solid particles larger than 8 μm are retained.
  - DHC from 1.5-8 kg of dirt depends on the filter size.
  - (Efficiency ↑ + DHC ↑) → Flow ↓ → that filter commonly used offline.

Before   After

**Fig. 12.13- Example of Stack Disc Elements
(Courtesy of C.C. Jensen Inc.)**

28

28

12.4.3.7- Filter Flow Rate

- High flow rate → physically larger size filters.
- ΔP should be < specified (at intended flow and maximum fluid viscosity).
- **Note:** differential cylinders → flow generation.

12.4.3.8- Filter Capacity versus Efficiency

- Normal size filter → is not designed to capture large quantities of dirt.
- Large amount of dirt → flushing or offline filtration using large filters.
- Highly restrictive media → better efficiency → small DHC.
- Less restrictive media → lower efficiency → large DHC.
- Better filter selection → balance between DHC and efficiency.

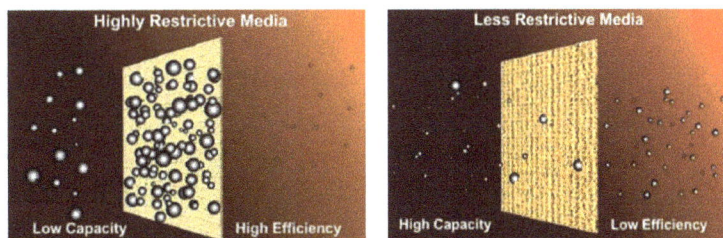

Fig. 12.14- Filter Efficiency vs. DHC (Courtesy of Parker)

29

29

12.4.4- ISO 3968: Differential Pressure Test

Test Purpose: Is to determine ΔP across the Filter.

ΔP depends on:

- Construction of the filter housing.
- Construction and type of filter element.
- Filter size and flow rate through the filter.
- Viscosity and specific gravity (SG) of the fluid flowing through the filter.

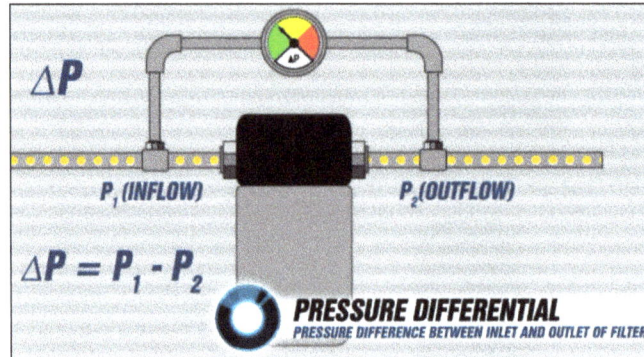

ΔP

P_1 (INFLOW) P_2 (OUTFLOW)

$\Delta P = P_1 - P_2$

PRESSURE DIFFERENTIAL
PRESSURE DIFFERENCE BETWEEN INLET AND OUTLET OF FILTER

**Fig. 12.15- Typical Filter Differential Pressure Test Setup
(Courtesy of Noria Corporation)**

30

30

$$\Delta p_{total} = (\Delta p_H + \Delta p_E) \qquad\qquad 12.4$$

Where, for a specific filter size, fluid flow, viscosity, and specific gravity:

- Δp_{total} is the total differential pressure across the filter assembly.

- Δp_H is the differential pressure across the filter housing (corrected based on SG)

- Δp_E is the differential pressure across the filter element (corrected based on SG & viscosity)

31

31

A typical flow-pressure drop curve for:
(a specific filter size, a specific clean filter media, and a specific fluid.

However, worst-case viscosity (at coldest operating T) should still be considered

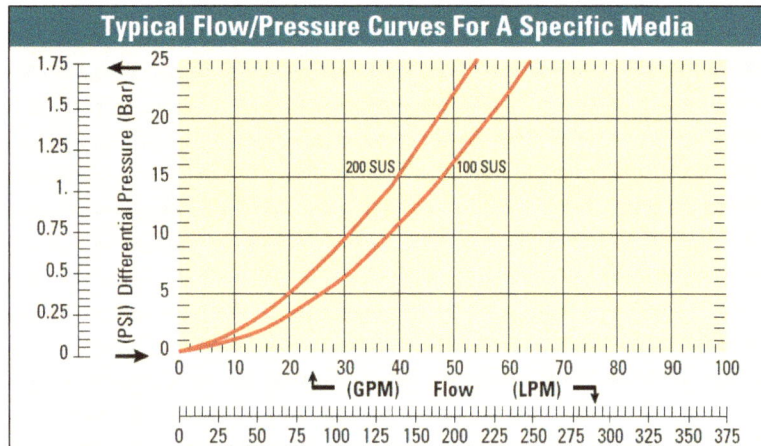

Typical Flow/Pressure Curves For A Specific Media

Fig. 12.16- Typical Flow-Pressure Curve for a Specific Filter (Courtesy of Parker)

32

Example 1 (Ref. Donaldson):

Given Data:

- Filter Data Sheet for a spin on filter (5 µm).
- Test fluid viscosity = 32cSt [150 SSU] at 100°F (37.7°C).
- Test fluid specific gravity = 0.9 at 100°F (37.7°C).

Exercise:

Find the filter head pressure drop for an actual hydraulic oil of 64 cSt viscosity and 1.1 specific gravity. Estimated flow rate is 150 gpm.

Solution:

$$\Delta p_{\text{Fiter Head}} = 3 \times \frac{64}{32} \times \frac{1.1}{0.9}$$

$$= 7.33 \text{ psid}$$

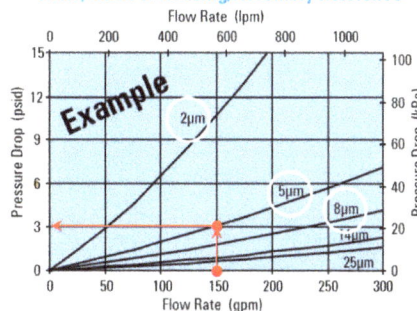

Filter Correction Calculation

$$\Delta P \text{ Filter} = \Delta P \text{ from graph} \times \frac{\text{New Saybolt Seconds Universal Viscosity (SSU)}}{150} \times \frac{\text{New Specific Gravity (S.G.)}}{.90}$$

- or -

$$\Delta P \text{ Filter} = \Delta P \text{ from graph} \times \frac{\text{New Centistokes Viscosity (cSt)}}{32} \times \frac{\text{New Specific Gravity (S.G.)}}{.90}$$

Clean Filter Assembly Pressure Drop (ΔP) Calculation

$$\Delta P \text{ Clean Filter Assembly} = \Delta P \text{ head} + \Delta P \text{ filter}$$

Filter, Head or Housing/Assembly Reference

Fig. 12.17- Example of Pressure Drop Calculation (Courtesy of Donaldson)

33

Example 2 (Ref. Schroeder):

Given Data:

- Filter Data Sheet.
- Test fluid viscosity = 32cSt [150 SSU] at 100°F (37.7°C).

Exercise:

For a filter NZ25-1N series, find the filter assembly total pressure drop for an actual hydraulic oil of 44 cSt (200 SUS) and 0.86 specific gravity. Estimated flow rate is 15 gpm.

Solution: See the figure.

$\Delta P_{housing}$

NF30 $\Delta P_{housing}$ for fluids with sp gr = 0.86:

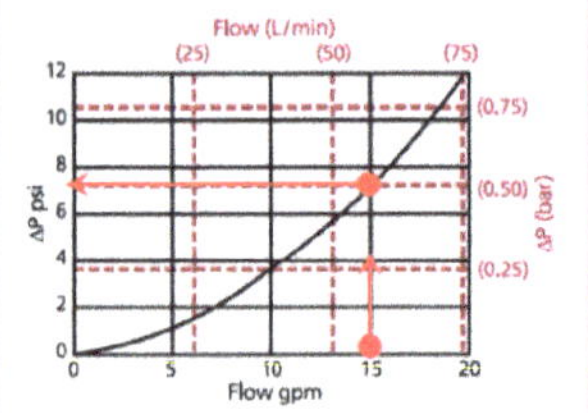

$\Delta P_{element}$

$\Delta P_{element}$ = flow x element ΔP factor x viscosity factor

El. ΔP factors @ 150 SUS (32 cSt):

	1N		1NN
N3	1.10	NN3	.77
N10	.17	NN10	.13
N25	.10	NN25	.07
NZ1	1.43	NNZ1	1.23
NZ3/NAS3	.92	NNZ3/NNAS3	.56
NZ5/NAS5	.71	NNZ5/NNAS5	.46
NZ10/NAS10	.57	NNZ10/NNAS10	.35
NZ25	.36	NNZ25	.20
		NNZX3	1.00
		NNZX10	.52

If working in units of bars & L/min, divide above factor by 54.9.

Viscosity factor: Divide viscosity by 150 SUS (32 cSt).

$\Delta P_{filter} = \Delta P_{housing} + \Delta P_{element}$

Exercise:

Determine ΔP at 15 gpm (57 L/min) for NF301NZ25SMS5 using 200 SUS (44 cSt) fluid.

Solution:

$\Delta P_{housing}$ = 7.0 psi [.50 bar]

$\Delta P_{element}$ = 15 x .36 x (200÷150) = 7.2 psi
or
= [57 x (.36÷54.9) x (44÷32) = .51 bar]

ΔP_{total} = 7.0 + 7.2 = 14.2 psi
or
= [.50 + .51 = 1.01 bar]

Fig. 12.18- Example of Pressure Drop Calculation (Courtesy of Schroeder)

34

Example 3 (Ref. Hydac):

EXAMPLE - an application with the following criteria would be sized as shown.

Conditions:
Fluid – Hydraulic Oil (ISO-32)
Specific Gravity – 0.86
Viscosity – 141 SSU
Flow Rate – 30 GPM
Fluid Temperature - 104°F normal

Filter Type Selected - Pressure Filter
HYDAC Model No. DF ON 240 TE 10 D 1.0 / 12 V -B6

HOUSING

ΔP Housing = ΔP Calculation *(From Curve in catalog)* x $\frac{\text{Actual Specific Gravity}}{0.86}$

ΔP Housing = 1.5 psid x $\frac{0.86}{0.86}$ = 1.5 psid

ELEMENT

ΔP Clean Element = ΔP Calculation x $\frac{\text{Actual Specific Gravity}}{0.86}$ x $\frac{\text{Actual Viscosity}}{141 \text{ SSU}}$

ΔP Clean Element = 30 GPM x 0.175 x $\frac{0.86}{0.86}$ x $\frac{141 \text{ SSU}}{141 \text{ SSU}}$

ΔP Clean Element = 5.25 x 1 x 1 = 5.25 psid

FILTER ASSEMBLY

ΔP Filter Assembly = ΔP Housing + ΔP Clean Element
1.5 psid + 5.25 psid = 6.75 psid

Fig. 12.19- Example of Pressure Drop Calculation (Courtesy of Hydac)

35

Example 4 (Ref. Pall):

Given Data:

- Filter Data Sheet shown in Fig. 12.20.
- Test fluid viscosity = 32cSt [150 SSU] at 100°F (37.7°C),
- Test fluid specific gravity = 0.9 at 100°F (37.7°C).
- Fluid flow = 100 l/min.

Exercise: Find the filter assembly pressure drop for a Series UH210 housing with -20 port sizes housing and an AN grade element of 13" length. Actual hydraulic fluid used has 50 cSt and specific gravity of 1.2. Estimated flow rate is 100 l/min.

Solution: see the figure.

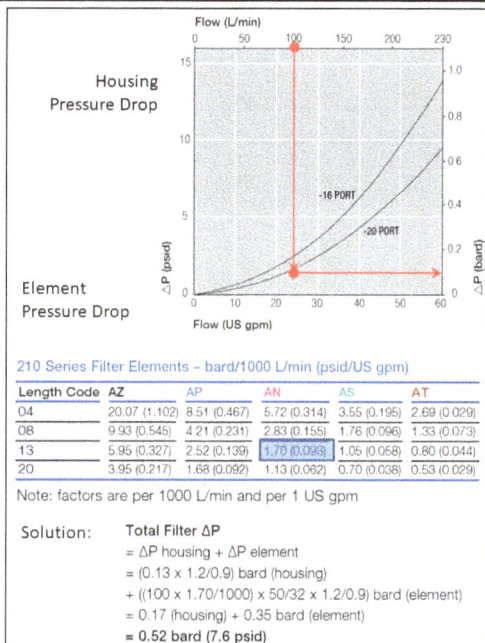

210 Series Filter Elements – bard/1000 L/min (psid/US gpm)

Length Code	AZ	AP	AN	AS	AT
04	20.07 (1.102)	8.51 (0.467)	5.72 (0.314)	3.55 (0.195)	2.69 (0.029)
08	9.93 (0.545)	4.21 (0.231)	2.83 (0.155)	1.76 (0.096)	1.33 (0.073)
13	5.95 (0.327)	2.52 (0.139)	1.70 (0.093)	1.05 (0.058)	0.80 (0.044)
20	3.95 (0.217)	1.68 (0.092)	1.13 (0.062)	0.70 (0.038)	0.53 (0.029)

Note: factors are per 1000 L/min and per 1 US gpm

Solution: Total Filter ΔP
= ΔP housing + ΔP element
= (0.13 x 1.2/0.9) bard (housing)
+ ((100 x 1.70/1000) x 50/32 x 1.2/0.9) bard (element)
= 0.17 (housing) + 0.35 bard (element)
= 0.52 bard (7.6 psid)

Fig. 12.20- Example of Pressure Drop Calculation (Courtesy of Pall) 36

36

12.4.5- NFPA (T-2.6.1): Rated Burst Pressure (RBP) of a Filter Housing

Definition: the static pressure at which <u>filter housing</u> structural failure occurs.

12.4.6- ISO 1077-1: Rated Fatigue Pressure (RFP) of a Filter Housing

Definition: the maximum allowable pressure for a <u>filter housing.</u>
Safety Factor = 4-6

$$RFP = \frac{RBP}{Safety\ Fator} \qquad 12.5$$

37

12.4.7- Cyclic Test Pressure (CTP) of a Filter Housing

Test Purpose: Some machines perform duty cycle (e.g. injection molding and die casting) → such repetitive functions requires (CTP) for filters.

Definition: Maximum pressure applied for certain number of cycles (typically 1 million cycles) before housing failure occurs.

Test Method: CTP is tested in accordance with standard test.

Mathematical: factor K that is obtained from tables associated with the above-mentioned standard based on confidence, assurance levels, materials of construction, and number of units tested.

$$CTP = RFP \times K \qquad\qquad 12.6$$

Example:
- RBP = 20,000 psi.
- Safety Factor = 4 →
- RFP = 20,000/4= 5,000 psi.
- K = 1.1 – 1.4 →
- CTP = 5,000 X 1.5 = 7,500 PSI

38

38

12.4.8- ISO 2941: Collapse Pressure of a Filter Element

Tet Purpose:
- Normal Operation → differential pressure indicates the filter status.
- Plugged filter OR Cold Start → filter element must withstand ΔP.

Definition: ΔP at which a structure failure of filter element and/or center tube occurs.

Test Method: Collapse pressure is determined by ISO 2941 standards.

39

39

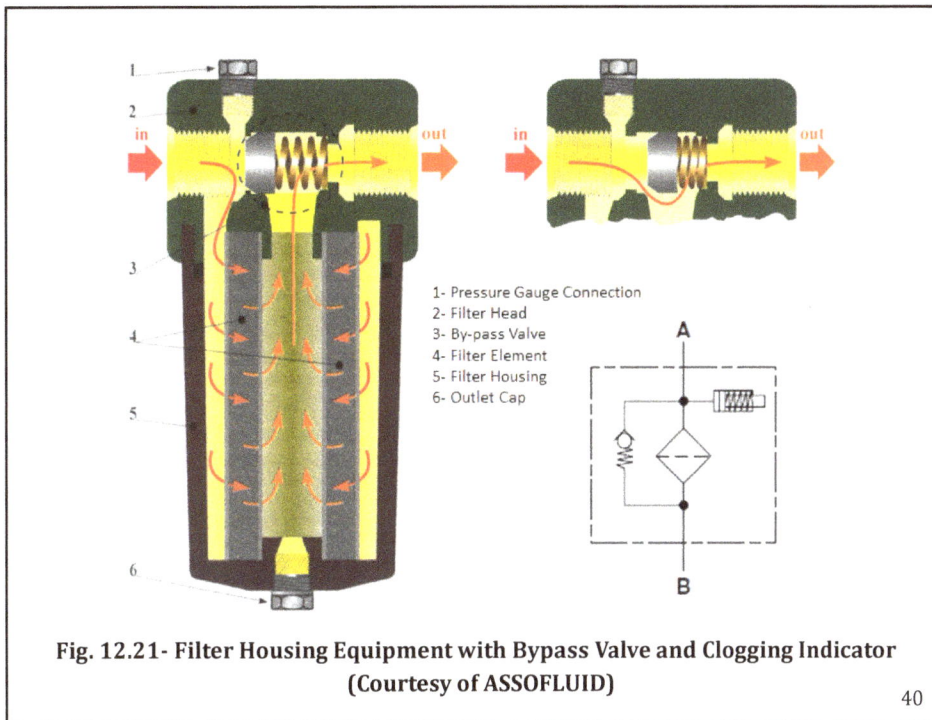

1- Pressure Gauge Connection
2- Filter Head
3- By-pass Valve
4- Filter Element
5- Filter Housing
6- Outlet Cap

Fig. 12.21- Filter Housing Equipment with Bypass Valve and Clogging Indicator (Courtesy of ASSOFLUID)

40

40

Rules of Thumbs:

- Total ΔP < half the bypass valve setting.

- Collapse Pressure >= 2 bypass valve setting.

- Servo and Proportional Valves $\rightarrow$

- No bypass valve in filter is recommended $\rightarrow$

- when the filter is clogged, all downstream functions are inoperative $\rightarrow$

- Collapse pressure PRV setting.

41

41

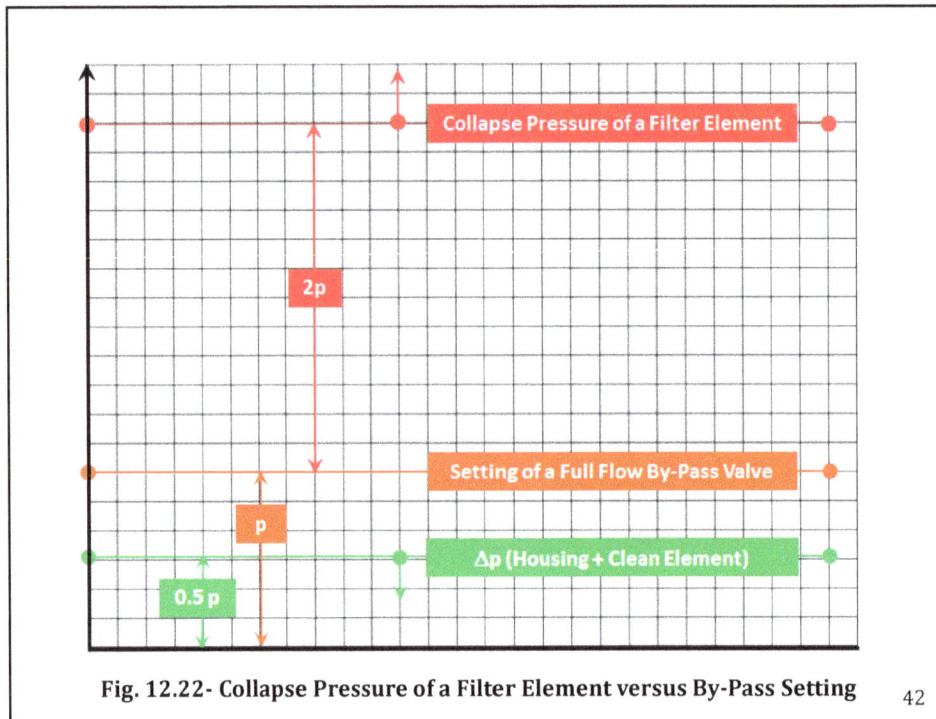

Fig. 12.22- Collapse Pressure of a Filter Element versus By-Pass Setting

42

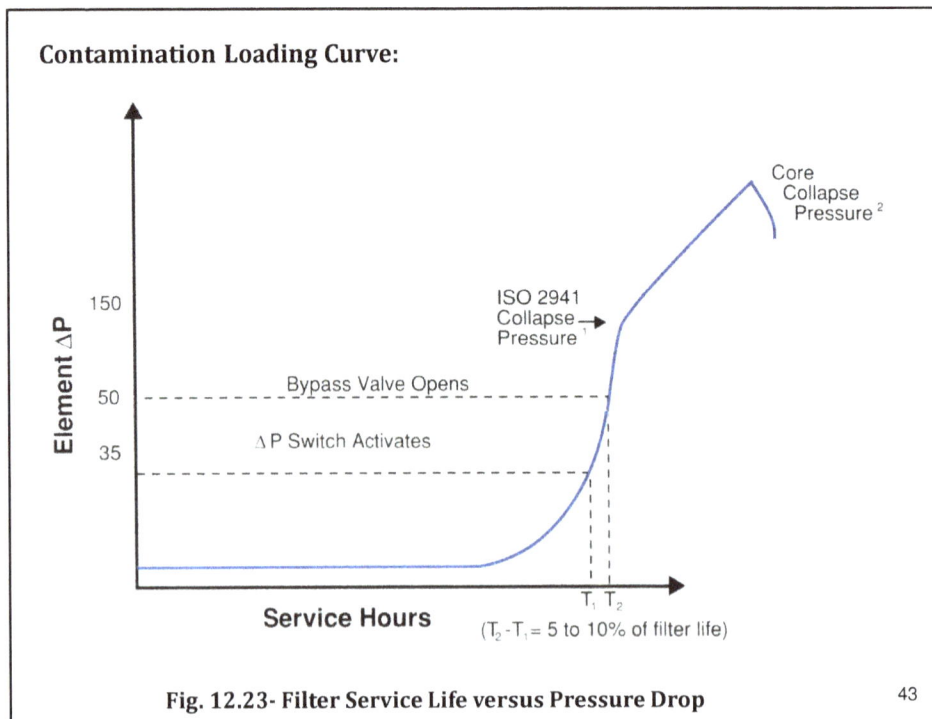

Contamination Loading Curve:

Fig. 12.23- Filter Service Life versus Pressure Drop

43

12.4.9- ISO 3723-2015: End Load Test

Test Purpose:

- To measure the ability of a filter element to resist axial deformation caused by differential pressure.

Test Method:

- Weights or other external load are used to simulate an axial force.

- When a filter element is subjected to the specified maximum axial load without permanent deformation, structural damage, or seal failure Filter

- → element is considered "PASS".

44

44

12.4.10- ISO 3724 OR ISO 23181: Flow Fatigue Test for Filter Element

Test Purpose:

- Some machines perform duty cycle (e.g. injection molding and die casting) → such repetitive functions requires Flow Fatigue Test.

- The test is used to predict the ability of a filter element to withstand structural failure due to flexing of the pleats caused by cyclic flow.

Test Method:

- The filter element is contaminated to its terminal differential pressure.

- The filter element is then subjected to a cyclic flow (0 – max).

- Number of cycles is prescribed by the element manufacturer,

- Usually based on (10-200) thousand cycles.

- No visual evidence of structural, seal or filter medium failure →

- element is considered "PASS".

45

45

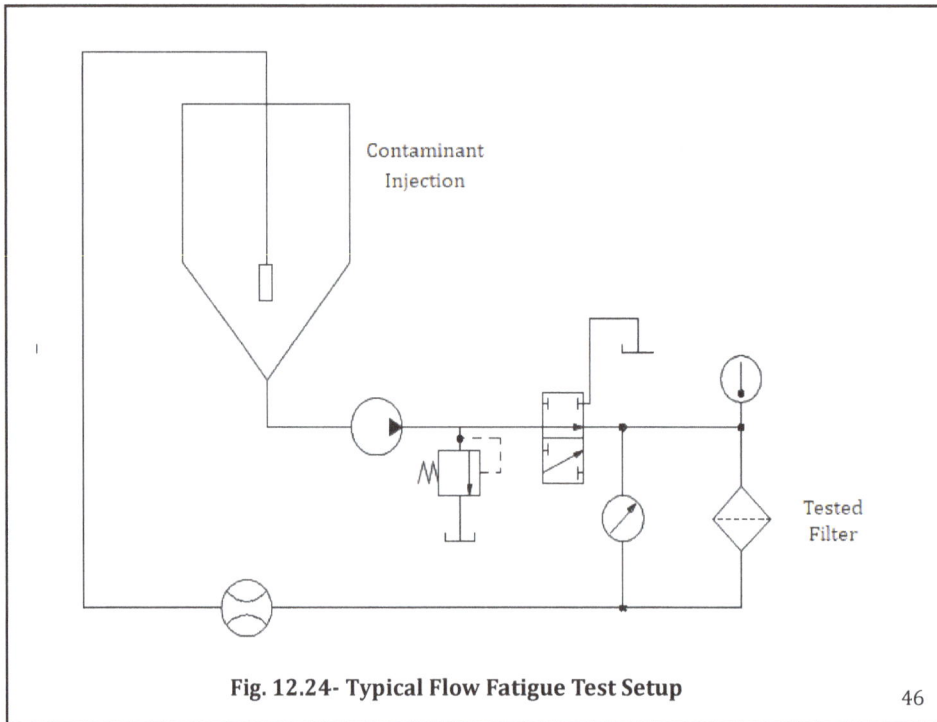

Fig. 12.24- Typical Flow Fatigue Test Setup

46

46

12.4.11- Filter Tests Pictogram

Video 466 (4 min)

Interpretation:

1. Dirt Holding Capacity and Beta Ratio "Multipass Test".
2. Filter Element Collapse Pressure.
3. Filter Element Differential Pressure.
4. Filter Element Structural Integrity "Bubble Point Test".

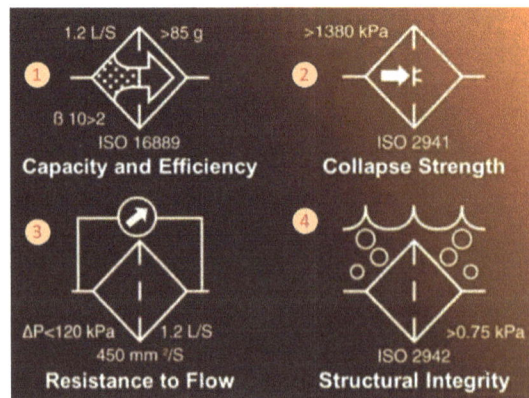

Fig. 12.25- Hydraulic Filters Tests Pictograms (Courtesy from Parker)   47

47

12.5-BP-Filters-05-Transportation and Storage

As reported by and excerpted from Donaldson Service Manuals:

Storage:
- Store a filter in a box or totally sealed from outside contaminant.
- Filters are sitting directly on metal shelves → condensation form on filters → wooden or plastic shelves are recommended.
- Make sure labels with product information and manufacturing dates are visible to personnel selecting from the shelves.

Shipping:
- Ship filters carefully (don't let them roll in the back of a truck).

Handling:
- Practice "first-in, first-out" with your inventory.
- Take care that the contaminant doesn't get on the new filter when remove it from the box.
- Handle filters carefully to avoid filter damage.

Video 422 (8 min) – Example of Filter Maintenance   48

48

Chapter 12 Reviews

1. Multipass test is aiming to measure?
   A. Maximum pressure can a filter withstand before structural damage.
   B. Maximum pressure can a filter withstand for a certain number of cycles before structural damage.
   C. Differential pressure across the valve.
   D. Filter efficiency.

2. Collapse Pressure is defined as?
   A. Maximum pressure can a filter withstand before structural damage.
   B. Maximum pressure can a filter withstand for a certain number of cycles before structural damage.
   C. Differential pressure across the valve.
   D. Filter efficiency.

3. Which of the following statement is TRUE?
   A. Collapse pressure = bypass valve setting
   B. Collapse pressure = pressure relief valve setting
   C. Collapse pressure < 0.5 bypass valve setting.
   D. Collapse pressure >= 2 bypass valve setting.

4. Total pressure drop across the valve should be?
   A. Collapse pressure = bypass valve setting
   B. Collapse pressure = pressure relief valve setting
   C. Collapse pressure < 0.5 bypass valve setting.
   D. Collapse pressure >= 2 bypass valve setting.

5. When a filter is rated as absolute for 5 micron size, this means that:
   A. Beta ratio = 75.
   B. Filter efficiency = 98.7.
   C. The filter retains 98.7% of the particles lager than and including 5 microns size.
   D. All the above is true.

Chapter 12 Assignment

Student Name: --- Student ID: -------------------

Date: --- Score: ------------------------

Questions: Explain test purpose and test method of Multipass standard test.

Answer:

Answers to Chapters Reviews

Chapter 1:

1	2	3	4	5	6	7	8	9	10
D	D	B	B	C	A	D	D	D	C

Chapter 2:

1	2	3	4	5	6	7	8	9	10
C	D	A	C	C					

Chapter 3:

1	2	3	4	5	6	7	8	9	10
A	B	C	B	A					

Chapter 4:

1	2	3	4	5	6	7	8	9	10
D	A	B	B	A	C	A	B	A	B

Chapter 5:

1	2	3	4	5	6	7	8	9	10
D	B	C	A	C					

Chapter 6:

1	2	3	4	5	6	7	8	9	10
C	D	A	D	A					

Chapter 7:

1	2	3	4	5	6	7	8	9	10
A	B	D	D	A					

Chapter 8:

1	2	3	4	5	6	7	8	9	10
B	D	B	C	C					

Chapter 9:

1	2	3	4	5	6	7	8	9	10
D	A	D	A	C					

Chapter 10:

1	2	3	4	5	6	7	8	9	10
D	C	B	A	D					

Chapter 11:

1	2	3	4	5	6	7	8	9	10
A	D	D	C	C					

Chapter 12:

1	2	3	4	5	6	7	8	9	10
D	A	D	C	D